DATE DUE

18 OCT 2002		
GAYLORD		PRINTED IN U.S.A.

Rapid Detection Assays for Food and Water

Rapid Detection Assays for Food and Water

Edited by

S. Clark
Microgen Bioproducts Ltd., Camberley, UK

K.C. Thompson
ALcontrol UK, Rotherham, UK

C.W. Keevil
University of Southampton, UK

M. Smith
Drinking Water Inspectorate, London, UK

ROYAL SOCIETY OF CHEMISTRY

The proceedings of the International Conference on Developments in Rapid Diagnostic Methods: Water and Food held at York on 15 – 17 March 1999.

Special Publication No. 272

ISBN 0-85404-779-4

A catalogue record for this book is available from the British Library

Published by The Royal Society of Chemistry,
Thomas Graham House, Science Park, Milton Road,
Cambridge CB4 0WF, UK

Registered Charity No. 207890

For further information see our web site at www.rsc.org

Printed by MPG Books Ltd, Bodmin, Cornwall, UK

Preface

This publication has arisen from a very successful international conference. The aim of the meeting was to present and analyse the evaluation of novel techniques to detect harmful chemicals and microbial pathogens in foods and waters. The various chapters concentrate on the specific topic areas of water microbiology, water chemistry, food microbiology and food chemistry. These four areas have both common and unique difficulties to overcome. In general the test systems under development and those being validated have to contend with pollutants at very low concentrations in large sample volumes and a wide range of sample matrices. This publication therefore represents the state of the art in highly sensitive chemical and biological detection systems for substances destined for the consumer. The techniques employed range from highly efficient concentration and sample preparation methods, through cell culture to end detection by a wide range of techniques. These end detection methods include microscopy, biosensors, colorimetry, immunoassay, molecular detection and molecular identification.

Contents

Rapid Methods in Water Microbiology

Rapid Methods in Water Chemistry

Rapid Methods in Food Microbiology

Rapid Methods in Food Chemistry

Rapid Methods in Water Microbiology

Professor C. William Keevil

Man cannot live without a wholesome supply of water for drinking, washing and preparing food. Despite rapid progress in science, the evolution of the built environment and improvements in sanitary practice, modern society continues to suffer from waterborne outbreaks of disease. Those in Western society should not be complacent because waterborne outbreaks can be larger there than those recorded in the Third World, a notable example being the outbreak of Cryptosporidiosis in Milwaukee, USA in 1993 which affected over 400,000 people. Consequently, the water and wastewater microbiology testing market runs into billions of dollars each year worldwide. Most testing is concerned with detection of faecal indicators of pollution, principally *Eshcherichia coli* and other faecal coliforms, as surrogates for the potential presence of dangerous pathogens such as *Salmonella enterica* or *Vibrio cholerae*. These tests and the use of chlorination over the past 100 years have caused a major improvement in human health and saved millions of lives each year.

The tests originally depended on classical culture techniques, with the stipulation that there should be no coliforms detected in say 100 ml of treated water, but now rapid enzyme-based chromagenic and fluorogenic tests have been approved for use in many countries. These tests have arisen due to the development of new technologies but have been driven by the factors of speed of test, cost of test, simplicity of test to be undertaken by relatively untrained staff, and large throughput of samples in each laboratory. Several of these factors are particularly important in the use of such tests in the Third World. These factors also relate to the drive to automate tests wherever possible, increasing speed and saving on manpower and sometimes reagent costs. Nevertheless, it should not be overlooked that a test must be fit for purpose and have the usual requirements of sensitivity, i.e. a very low limit of detection but sometimes taken to mean that there are no false negatives, and specificity (i.e. no false positives). Every test must go through a vigorous validation procedure and receive regulatory approval from organisations such as ISO and CEN.

The wastewater treatment industry has been relatively slow to adopt the newer technologies developed in the clinical, veterinary, food and potable water industries. However, there has been recent concern about new and re-emerging pathogens being transmitted by faecal contamination and the risk of their presence in wastes when recycled to agricultural land. This has caused a reappraisal of what tests are appropriate for stored and/or treated human and animal wastes. In the USA, the EPA Part 503 Sludge Regulation now stipulates that a sludge is acceptable for use on land if it contains less than 1000 faecal coliforms per g dry weight of sludge and less than 3 salmonellae, 1 virus and 1 viable helminth per 4 g (Federal Register, 1993). These stringent regulations have highlighted the need for better methods of detection for each type of microorganism in sludge. Moreover, there is a growing realisation that stored and treated wastes may contain sub-lethally damaged pathogens with the possibility of them still being able to cause disease. Consequently, newer culture techniques are addressing the requirement of including a resuscitation, pre-enrichment step before enumeration on agar media or detection with molecular and immunological technologies. This extra step inevitably slows down the detection method, making it less than rapid and probably more expensive. Perhaps the future holds a compromise of using simple tests, such as immunological dip sticks, for rapid on-site monitoring for a pathogen's presence, followed by a subsequent full laboratory resuscitation analysis of the presumptive positives to look for the viable organism. This approach may be criticised however because of the insensitivity of immunological techniques to detect low numbers of the target organism. This is certainly one area where the diagnostics industry should look to increase the sensitivity of detection. Virus detection has become easier and rapid with the development of PCR, RT-PCR and immunological detection technologies. However, it is unlikely that the assessment of viability will become rapid or cheap because of the need for cell culture-based assays with the caveat of their sensitivity to toxic inhibitors in the wastes. As a consequence, rapid methods are unlikely to dominate the water testing market in the near future but useful advances will be made by integrating some of their technology into more tedious assay protocols.

SUITABILITY OF MICROBIAL ASSAYS FOR POTABLE WATER AND WASTEWATER APPLIED TO LAND

C. W. Keevil

School of Biological Sciences, University of Southampton, Bassett Crescent East, Southampton SO16 7PX, UK

1 INTRODUCTION

Emerging methods of sample preparation and pathogen detection have been developed in the food, clinical and veterinary industries, involving direct specific detection or post-enrichment culture screening with molecular, immunological and enzyme assay technologies. Some of these have also been used, with some success, by water companies involved in the supply of potable water and have been validated as ISO and CEN standards. Specific immunomagnetic separation technologies, avoiding selective enrichment, appear particularly promising and are already incorporated into several national standards for *Cryptosporidium* and *Giardia* detection in raw and treated waters. The human population continues to expand at an ever increasing rate, requiring greater intensification of animal rearing for food production, and there are inevitably greater demands on the environment to deal with the human and animal faecal wastes which are generated. Many instances of foodborne and waterborne disease outbreaks point to the transmission of new and re-emerging pathogens in faecal wastes which are recycled to land for economic and environmental reasons. The challenge for analysts, therefore, is to adapt some of the methods described above, or develop new methods, to provide an evaluation of safe storage and treatment practices for pathogen removal before the wastes can be safely recycled to land. This problem has been exacerbated by the realisation that some pathogens may be sub-lethally damaged, but possibly still capable of causing disease, due to the storage and treatment processes and might be missed using conventional culture recovery techniques. The molecular techniques are presently more suited to pathogen detection, not viability, and research is now focusing on appropriate resuscitation techniques involving either presence/absence or the newer quantitative methodologies.

This paper will address some of the current legislation to ensure safe pathogen limits in wastes and the practicality of available or prototype methods for their detection in complex faecal matrices.

2 MICROBIAL ASSAYS FOR POTABLE WATER

The water industry has relied for many years on surrogate general indicators of faecal pollution to maintain a wholesome supply of potable water. Bile salt-tolerant, facultatively anaerobic bacteria isolated on MacConkey agar at 37°C, and capable of producing acid and gas from lactose at this temperature, were described as "coliform organisms". Semi-selective MacConkey medium (supplied commercially by Oxoid, Difco *et al.*) was devised to reduce the overgrowth when isolating faecal lactose-fermenting microorganisms; the medium incorporates sodium taurocholate (bile salts), to inhibit common environmental contaminants, and lactose is fermented to produce the neutral red coloration. Pale colonies of *Salmonella* or *Shigella* spp. are usually seen among red colonies of *E. coli*. Because coliform organisms can be of non-faecal origin, presumptive faecal coliforms were described as those capable of producing acid at 44°C, the majority subsequently being identified as *Escherichia coli*. Total coliform detection came to be relied upon because of the relative ease, speed and low cost of detection.

Alternative media include membrane lauryl sulphate broth and agar (MLSB; MLSA), violet red bile agar (VRBA), lauryl tryptose lactose agar (LTLA), desoxycholate agar and CLED agar. MLSA and desoxycholate agar (DCA) are able to differentiate *E. coli* from *Klebsiella oxytoca* in environmental water samples by virtue of *E. coli*'s ability to grow at 44°C as well as 37°C.[1] MLSB is recommended by the PHLS/SCA[2] for the isolation and enumeration of total coliforms and *E. coli* by membrane filtration. The recommended method for membrane filtration involves incubation of duplicate membranes on MLSB-soaked pads at 30°C for 4 h, followed by incubation of one at 37°C for 14 h (for presumptive coliforms) and the other at 44°C for 14 h (for presumptive *E. coli*).[3] Minerals Modified Glutamate Medium (MMGM) is recommended for the isolation of coliforms from waters by multiple tube and is slightly superior to LTLA.[4] MT7 agar (Difco) was developed to isolate injured coliforms and is particularly suited to the recovery of chlorine treated samples.[5]

Latterly, enzyme detection methods have been developed, based on the hydrolysis by β-galactosidase of chromogenic and fluorogenic substrates, such as 5-bromo-4-chloro-3-indolyl-β-D-galactopyranoside (X-gal) and β-methylumbelliferyl-β-D-galactopyranoside, respectively. These methods offer a much quicker indication of the presence of "biologically active" coliforms. Furthermore, specificity to detect *E. coli* has been introduced by incorporation in various media of chromogenic and fluorogenic substrates of β-glucuronidase. These assays can be performed in liquid culture or aid in the differentiation and identification of bacterial colonies recovered on agar media. Incorporation of 4-methylumbelliferyl-β-D-glucuronide (MUG) into any of these media may give some indication of the presence of *E. coli* due to β-glucuronidase activity producing blue-white fluorescent colonies when viewed under ultraviolet light at 366 nm.[6] Incorporation of MUG at 50 or 100 mg l⁻¹ for liquid or agar media, respectively, can be used to provide presumptive evidence for the presence of *E. coli* which should be confirmed by further biochemical or immunological tests. Sartory and Howard[7] described an agar medium, membrane lactose glucuronide agar (mLGA), for the simultaneous detection of *E. coli* and coliforms by membrane filtration without the need for UV light observation. The 5-bromo-4-chloro-3-indolyl-β-glucuronide chromophore in mLGA turned β-glucuronidase-producing colonies green (lac⁺ gluc⁻ colonies are yellow) and gave similar *E. coli* confirmation rates (92%) as isolation on MLSA.[8] Pyruvate was also added to mLGA to substantially improve recovery of chlorine stressed coliforms. This method has now been successfully adapted to quantify the numbers of *E. coli* surviving in raw and treated sewage sludge [9]; agar plates are incubated at 30°C for 4 ± 0.5 hours, followed by 44.0°C for 14 hours.

The use of bacterial indicators has demonstrated the effectiveness of physical and chemical treatment barriers at treatment works and in supply to prevent distribution of viable pathogens. There are several caveats, however. Are some chlorine-treated coliforms only sub-lethally damaged and, although viable, incapable of growth in the normal isolation media? As discussed, several chlorine-stress recovery agar media have been developed to help overcome this problem. Are there pathogens more tolerant of chlorine than the coliforms? Occasional waterborne outbreaks of infection caused by pathogens resistant to chlorine disinfectants, particularly *Cryptosporidium parvum*, have highlighted the need for more specific tests of pathogens in treated water. Because of the potentially low infectious dose of some of these emergent pathogens, the assay methods require a high precision to provide the confidence of being able to detect very low numbers in large volumes of treated water. This volume provides several important examples of the progress made in this area.

3 RECYCLING OF WASTES TO LAND

In 1992 approximately 470,000 dry tonnes of sewage sludge or biosolids were disposed of to soil in the UK. This amount has now doubled, due mainly to the ban on sea dumping of sewage which came into force in 1998 as a result of the EC Urban Waste Water Treatment Directive (1991; incorporated into UK law 1994). Land application is recognised as the Best Practicable Environmental Option for using sewage sludge. Sewage sludge and agricultural wastes are recycled to soil with the aim of improving soil condition and fertility.[10] However, sewage sludge may contain a range of microorganisms pathogenic to man, including bacteria (e.g. *Salmonella, Campylobacter, Listeria* and various strains of *E. coli*), virus particles (e.g. Polio and Hepatitis), protozoa (e.g. *Cryptosporidium*) and other intestinal parasites (e.g. Helminths) (Table 1).

Table 1 *Principle microorganisms in organic wastes pathogenic to animals and man*

Genera	*Principal species or strain*
Escherichia	*coli* (O157:H7; O26, O103, O111, O145)
Salmonella	*enteritidis* PT4, *typhimurium* DT104
Shigella	*dysenteriae, sonnei*
Campylobacter	*jejuni, coli*
Listeria	*monocytogenes*
Cryptosporidium	*parvum* (genotype 1 and 2)
Giardia	*lamblia*
Cyclospora	*cayetanensis*

Helminths

Without suitable treatment, there is potential for pathogens present to wash into adjacent surface waters, contaminate crops (fresh produce is of particular concern), or spread directly to man or farm and domestic animals using the land. Application of human sewage sludge currently represents a small proportion of waste applied to agricultural land; by far the greatest amounts are contributed by a variety of animal wastes including compost, faecal slurries, poultry litter, *etc.* (Table 2). The availability of suitable detection methods to facilitate the understanding the survival of potential human and animal pathogens in these wastes, before and after application to land, is critical to delivering safe agricultural products to the market place.

Table 2 Sludge, agricultural and food processing (e.g. abattoir) wastes recycled to land in UK

Waste	Amount recycled p.a.
Sewage sludge	>1m tonne post 1998
Cattle slurry	40m tonne
Cattle manure	31m tonne
Pig manure	5.4m tonne
Poultry manure	4.5m tonne
Abattoir waste	0.4m tonne

EU Council Directive 86/278/EEC regulates sewage sludge applications to agricultural land throughout the EU. The Directive is implemented in the UK by statutory instrument The Sludge (Use in Agriculture) Regulations 1989 (SI 1989, No 1263, as amended SI 880 1990; HMSO), regarding the application of raw and treated sewage sludge to agricultural land. The regulations are supported by the DOE Code of Practice for Agricultural Use of Sewage Sludge 1989 (revised 1996; DOE). These strictly limit how the sewage sludge is to be applied, under what conditions and to which crops, reducing as far as possible occasional contact with animals and man. The Code of Practice was developed in the 1970s from the data available at the time, and before the emergence of highly infectious pathogens such as *E. coli* O157.

However, recent articles in the press have indicated the increasing public concern about the dangers associated with the routine dumping of blood, offal and raw sewage onto British farmland (see, for example, the "Fields of filth" article by Day[11]). These articles paint an alarming picture of the perceived risks, largely because of concerns over newly emergent foodborne diseases believed to be caused by variant prion proteins, bacteria such as verocytotoxic *E. coli*, *S. enteritidis* and antibiotic-resistant *S. typhimurium* DT 104, and parasitic protozoa such as *C. parvum*. The problems these now cause have exacerbated the recurring problems of foodborne *Campylobacter*, *Salmonella* and *Shigella* spp. There are over 100,000 cases of gastro-enteritis reported in the UK each year, with some believing that the actual number infected is close to 1 million.

Pathogens such as toxin-producing *Escherichia coli* O157 may indeed be occasionally present in sewage and this strain's very low infectious dose (possibly <10 cells) has heightened public health concern over its transmission and persistence in the environment.[12] Work in strictly controlled laboratory conditions has shown that this pathogen is physiologically versatile (James and Keevil, 1999) and may survive for months in cattle faeces, manure, model soil systems (Figure 1) and river water, highlighting the potential to contaminate growing crops.[13, 14] Nevertheless, caution must be urged in interpretation of the data at this time as there has yet to be an investigation under field conditions of pathogen survival in a representative range of the different untreated and treated sewage or abattoir waste which may be applied to land. Preliminary work in the laboratory indicates that *E. coli* O157 may be killed in sewage sludge during lime stabilisation (Maule and Keevil, unpublished data).

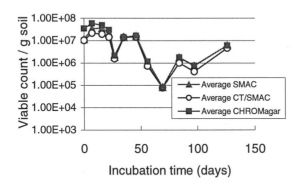

Figure 1 *Survival of* E. coli *O157:H7 in grassed soil cores at 18°C*

Similarly, *C. parvum* oocysts and *C. jejuni* may persist in aquatic environments for several months, attached to surfaces[15-17], and have been associated with foodborne outbreaks.[18] Salmonellae cause a significant number of foodborne infections associated with fresh produce, including *S. enteritidis* PT4, and there is concern over the increasing disease incidence of the multiple drug resistant *S. typhimurium* DT104 which is found in cattle and poultry waste. *L. monocytogenes* is widely distributed in nature and work has shown it to be a versatile, robust pathogen capable of survival and growth at low temperatures whilst maintaining a virulent phenotype.[19] It is undoubtedly food borne, however it cannot be truly described as a food poisoning organism since its symptoms are rarely associated with the gastrointestinal tract.[18] Infection ranges from symptomless carriage to fatal septicaemia, meningitis or neonatal abortion. In addition to consumption of dairy products, other implicated foods include meat pates, coleslaw, poultry, fin fish and shellfish. Several large foodborne outbreaks occurred in North America in the 1980s which were notable for a case fatality rate of 30%.[20]

In addition to domestic sources, effluents from agricultural and commercial enterprises (i.e. cattle markets, abattoirs etc.) can be responsible for releasing large numbers of *Cryptosporidium* sp. oocysts into sewage. The first recorded outbreak of waterborne Cryptosporidiosis was associated with contamination of a well water supply with oocysts derived from sewage.[21] Examination of stool specimens from 79 residents who consumed water from the well source indicated that gastrointestinal illness was 12 times greater than in the surrounding communities which consumed water from a different water source. Since then, numerous waterborne outbreaks of human Cryptosporidiosis have been reported. The potential for causing disease in large numbers of individuals was demonstrated in the Milwaukee outbreak which affected an estimated 403,000 individuals.[22] Through the use of molecular techniques, four isolates from this outbreak have been found to infect only humans, indicating that human sewage might have been the source.[23]

Approximately 6000 cases of Cryptosporidiosis in England and Wales are reported to the PHLS each year, and this is believed to be gross underestimate of actual cases. There are more cases of Giardiasis reported, but these are mainly sporadic and the source of infection untraceable. The environmental reservoir for the aetiological agent, *Giardia intestinalis* (*lamblia*), is unclear but it has been reported to be present in beavers (hence the name "Beaver fever") and farm animals, and spread in water.[24] Although *Giardia* cysts are considered less robust than *Cryptosporidium* oocysts, one report has suggested that 10% of infective cysts can survive several weeks of incubation in swine manure slurry.[25] This reinforces the need for proper treatment of organic wastes before disposable to agricultural land.

The term enteroviruses is a general description for viruses that infect and replicate in cells of the gastrointestinal tract. A prominent family of viruses within the large group of viruses that are found in the intestinal tract are the Picornaviridae. Within this family are several genera including polioviruses, coxsackieviruses (two groups, A and B), echoviruses and enteroviruses (a specific name and as opposed to the general description above). Enteroviruses infecting humans are found world-wide and humans are the only known natural hosts. Young children are the most susceptible to infection. Transmission is usually by the faeco-oral or by the respiratory route where there is an associated respiratory illness. The viruses may be excreted in the stool for many weeks. Poliovirus in sewage is decreasing due to vaccination and the increased use of disposable nappies diverting contaminated faeces to solid waste. Enteroviruses have been detected in water, soil, vegetables and shellfish and may possibly be transmitted in the community by contact with contaminated food or water. In practice, nearly all foodborne viral illness is either hepatitis A or viral gastro-enteritis due frequently to small round structured viruses (SRSV).[18] Food borne transmission of other viruses causing gastro-enteritis, such as rotavirus, appears rare. Current detection methods still rely on the "gold standard" of electron microscopy, but newer methods involving cell culture (although this is not possible yet with SRSV) and/or PCR, RT-PCR and antibodies are being developed.[26] Cell culture at least allows the detection and possible quantification of viable organisms.

Given the potential problems of pathogens in faecal waste, poultry litter, slaughterhouse waste *etc.*, it is essential that before these biosolids are disposed to land they are given an appropriate treatment which will reduce any pathogen content to an acceptable level or stress them sufficiently that they will not persist in the environment or on crops. The regulations or guidelines for this treatment vary world-wide and are under review. There are currently no defined limits on the microbiological quality of sludge applied to land in the UK, only requirements for type of treatment and application. Recently, the UK retailers, via the British retail Consortium, have raised concern over the microbiological risks from applying sewage sludge to agricultural land. In response, ADAS has produced a "Safe" Sludge matrix which gives recommended minimum time periods between the application of sludges to land and the use of that land for food production.[10] However, there is minimal information available on the levels of pathogens in wastes and the survival and spread of these pathogens on the farm. Therefore, it is difficult to accurately assess the most appropriate no harvesting/grazing periods for different types of wastes. In the USA, the EPA Part 503 Sludge Regulation stipulates that a sludge is acceptable for use on land if it contains less than 1000 faecal coliforms per g dry weight of sludge and less than 3 salmonellae, 1 virus and 1 viable helminth per 4 g.[27] These stringent regulations have highlighted the need for better methods of detection for each type of microorganism in sludge.

It is becoming increasingly clear, therefore, that more research is required into the survival of pathogens in soil and the various matrices associated with human and animal wastes, not least because the technologies to detect low numbers of environmentally stressed pathogens have improved. Consequently, there is a need to revisit earlier work and provide the new data essential to ensure that the Code of Practise is still appropriate, and if necessary updated, to reassure the public that waste disposal and farming practises are safe to the environment and human health. To initiate this process, a short review was undertaken for the UK Water Industries Research (UKWIR) to assess methods of bacterial pathogen detection in untreated and treated biosolids and recommend the most appropriate for further comparison in the laboratory and in the field.[28] The review identified various culture and molecular methods which have been developed in the last few years to improve the detection of *E. coli* O157, *Salmonella* and *Campylobacter* spp., principally in food and clinical samples but also with some application to veterinary samples, including faeces. Some of the methods identified appear appropriate to detect pathogens in raw and treated sewage but have not been properly assessed nor validated. Consequently, on-going work is now capitalising on these studies to develop improved methods of detection for bacterial, viral and protozoal pathogens in sludge and animal wastes.

4 AVAILABLE TECHNOLOGIES

The concerns of food retailers and the general public regarding food hygiene make it clear that many thousands of untreated and treated biosolid samples recycled to land, as well as potentially contaminated foods, will need to be tested each year. This will demand mass screening techniques, ideally at low cost. Due to the low infectious dose of some of the pathogens, the rapid detection methods must be sensitive and specific, yet robust in the complex matrices involved. Different approaches are required for the type of monitoring required; for example, the Regulator may require accurate quantification of pathogen numbers for assessing process control of treatment methods, using specific culture, immunological and molecular techniques. However, where the efficiency of specific treatment effect has to be demonstrated (e.g. a 6-log kill) then the culture techniques are mandatory because the molecular approaches cannot truly discriminate between viable and dead cells. Conversely, it may be more suitable for an operator applying biosolids to land to rapidly screen biosolids for their biological activity and stability, using respiration, enzyme assay, dipsticks or most probable number techniques. Recent developments in these areas will be presented.

4.1 Summary of technologies and preparation time

The methods to be employed will ultimately depend on time, cost, and throughput of samples per day. The flow diagram (Figure 2) illustrates the possible integration of various isolation/detection technologies into protocols for each bacterial pathogen.

Figure 2 *Integration of possible isolation and detection techniques*

PRE-SCREEN/SAMPLE PREPARATION

	Preparation Time
General Screen (Oxygen uptake, Vital stains, Total coliforms)	5-120 min
Sample Preparation - (Stomacher, Pulsifier, Release buffer)	10 min

DIRECT SPECIFIC DETECTION (pre-broth)

Molecular or immuno detection	4 h
non-selective pre-enrichment (buffered peptone water etc.)	4-24 h
Molecular or immuno-detection	4 h
IMS and/or selective enrichment	1 h IMS 24 h enrichment
Impedance detection	6-12 h

POST BROTH SCREENING

Immunoscreen ----Manual - (Clearview, Rapitest, Tecra)	1-2 h
---- Automated - (Organon-Technica, Transia, Tecra, Rhone-Poulenc, Biocontrol)	1-2 h
---- Automated Magnetic - (FOSS Electric)	1-2 h
Molecular Screen ---- PCR, 16S/23S rRNA	3-4 h
Culture Confirmation ---- Latex, Biochemical reactions, Chromogenic media, (Closed screening, VIDAS)	10 min latex 24h biochemical
Biochemical Identification ---- (API, BIOLOG, VITEK, Micro-ID, Crystal, Microbact)	24 h

4.2 Pre-screen/sample preparation

The Royal Commission on Environmental Pollution's Report (1996) on the sustainable use of soil called for the UK Government to review policies on heavy metals and pathogens in sewage sludge and, more generally, the use of industrial wastes spread on agricultural land. Subsequently, the House of Commons Environment, Transport and Regional Affairs Committee published their Second Report entitled Sewage Treatment and Disposal (1998). A key recommendation is the requirement, by 2002, for all sewage sludge to be treated to tertiary level at all times and in all places using processes or combinations of processes that will help reduce nutrients as well as pathogens. Moreover, sewage sludge that is to be recycled to land should be subjected to stabilisation and pasteurisation. Sludge is normally assumed to be stabilised after 5 days thermophilic digestion, 30 days mesophilic digestion or after 3 months storage in an open lagoon.

The most obvious method of assessing stability is by measuring the respiratory activity of all of the microorganisms present. Pasteurisation is expected to kill or incapacitate the majority of the microorganisms present, producing a marked decrease in respiratory activity and oxygen uptake rates. This can be determined using one of the many protocols available for measuring oxygen uptake rates. Automated, robust respirometers have become available, such as the microprocessor controlled, manostatic respirometer (Merit[TM] Respirometer, Terra Nova Systems Ltd, York). This is currently being used to determine the inhibitory effects of chemical and industrial waste waters on the respiration of activated sludge.[29] Oxystatic respirometers measure the quantity of oxygen required to maintain a constant pressure in the vessel and possess several advantages over manostatic respirometers. The MERITOX[TM] oxystatic respirometer (AE Instruments, Brompton-on-Swale) has been performance tested for its suitability for use in a SCA Blue Book method.[30] The method measures the rate of oxygen consumption from a mixture of a synthetic sewage (peptone, meat extract, urea and salts) and activated sludge (6 g l^{-1} to provide a suitable ratio of nutrients to biomass) or other wastewater.

Alternatively, the respiratory activity of individual microbial cells can be assessed either manually with a microscope or using automated cell cytometry (see later). These methods rely on the transfer of electrons produced during respiratory activity to a chromophore such as 2-(p-iodophenyl)-3-(p-nitrophenyl)-5-phenyltetrazolium chloride (INT) or a fluorochrome such as cyanoditolyl tetrazolium chloride (CTC); in the reduced state these electron acceptors precipitate internally and exhibit a deep red colour or bright red fluorescence, respectively.[31, 32] The rate of formation is proportional to the respiratory activity. This is a measure of "vitality" rather than "viability" but may provide some evidence that although a cell appears unculturable on agar media, it might be stressed/sublethally damaged rather than dead.

The chromogenic and fluorogenic enzyme methods developed to detect coliforms in treated potable water may be applicable to the detection of "biologically active" coliforms in wastewater. Thus, it should be possible to monitor the general biological activity of untreated and treated wastes by assaying β-galactosidase activity. The enzyme is normally inducible and the assay will require the presence of lactose or non-metabolisable IPTG (isopropyl-β-D-thiogalactopyranoside) as inducers. Again, the process of induction requires biological activity.

4.3 Methods for sample preparation

A major consideration for the recovery and quantification of microorganisms from complex wastes is to ensure that they are completely released from any matrix to which they are attached, such as faecal debris, and adequately dispersed to prevent clumping of cells before pre-enrichment and/or culture. This is especially important if the sample is to be filtered from the debris before culture. For samples such as sewage sludge or animal litters it may be necessary to use a buffered releasing diluent such as 0.01% (w/v) Tween 80 or phosphate buffered manucol antifoam (PBMA): this contains 0.25% (v/v) manucol releasing agent and 0.01% (v/v) GE Silicone 60 to prevent foaming during homogenisation.

Homogenisation techniques have included the use of mortar and pestle, sonication, vortex agitation or Waring blenders. Probably the most widely used technique, however, has involved the Stomacher® Lab Blender (Seward Ltd, London) (Sharp and Jackson, 1972).[33] The apparatus works by a paddle action which repeatedly crushes samples in a plastic bag to achieve homogenisation. It is of robust construction and the latest versions have 3 preset speed and 3 time settings. This apparatus is particularly suited to samples where the microorganisms are deep inside cohesive structures.

The ideal dispersion technique is one which can remove pathogens from complex wastes without disintegrating the matrix which could lead to difficulties in recovery of the pathogen. A promising solution is the recent development of the Pulsifier™ (Microgen Bioproducts Ltd, Camberley), which uses a stainless steel ring to vibrate at 3500 rpm and beats the outside of a bag to produce a combined shock wave generator and intense stirring.[34] This drives attached microorganisms into suspension without disintegrating complex matrices such as foods or wastes. Samples of up to 100 ml or 250 ml volume are contained in sealed plastic bags, and the apparatus is compact for housing in a safety cabinet, making the technique particularly suited for the detection of ACDP Category 3 pathogens such as *E. coli* O157 in sludge and food contaminated with faeces (Bown and Keevil, unpublished data).

The Pulsifier sample release technology has already been proven to be as good or better than the Stomacher for the recovery of a range of bacteria from ground beef, fish, pat—, cheese, peas, sprouts, alfalfa, celery, carrots, oats, rice etc. The Pulsifier has the added advantage that it does not crush food and, therefore, leads to suspension of microorganisms with less food debris. This should be ideal for the release and subsequent recovery of pathogens in complex wastes.

For disperse samples and/or low numbers of the target organism it may be necessary to filter samples (e.g. 0.4 µm cellulose acetate filter or polycarbonate membrane) before incubation in a pre-enrichment broth. If there is a lot of particulate matter then a coarse pre-filter of spun glass or other material may be required to avoid clogging of the microbiological filter.

4.4 Molecular detection

Rapid advances in molecular biology and molecular taxonomy are making it clear that many microorganisms exist in the environment, including potable water and wastewater, which cannot presently be cultured. Research has indicated that only approximately 1% of the bacteria in potable water can be cultured [35], yet the rest appear metabolically active (Roszak and Colwell, 1987).[36] This non-culturability is presumed to be because:

- they are of previously unidentified genera/species whose physiology and growth requirements are not understood (implying that conventional laboratory growth media contain inappropriate nutrients), or
- they are environmentally stressed due to nutrient limitation, extremes of temperature, pH, redox, osmolarity etc., or to the presence of disinfectants such as chlorine, and may be termed viable but nonculturable (VNC) using routine laboratory media.

Knowledge of these non-culturable species is important because in the environment and the built environment they play a role in biofilm formation (causing biofouling, heat loss or corrosion of pipework, but making important contributions to wastewater treatment processes) and might also provide a shelter for bacteria causing infectious diseases.[16] In particular, sub-lethally damaged or VNC pathogens such as VTEC, *Salmonella, Shigella* or *Campylobacter* spp. may be present in untreated and treated wastes and, although possibly remaining capable of causing infection, are undetectable by routine culture.[36, 37]

Where microorganisms can be cultured from low nutrient environments on specialised media, such as low nutrient R2A media[38], there is frequent disagreement over their identity when characterised using commercial API, BIOLOG and VITEK biochemical databases (as discussed previously). Even then, they make take 7-10 days to grow before identification. However, 16S and 23S rRNA sequences provide a unique signature for each prokaryotic species. Phylogenetic analysis of the rRNA sequences can be used to identify recovered bacteria in relation to well characterised strains, or the creation of new genera (e.g. within the α, β, or γ subclasses of the Proteobacteria or Eubacteria). The development of *in situ* hybridisation with rRNA-targeted oligonucleotide probes [39] has allowed rapid identification of bacteria within their natural habitat. Furthermore, where species are non-culturable, strain specific rRNA probes can be produced (using conserved primers as original templates to amplify the variable regions for sequence analysis) to determine their abundance *in situ*.

The polymerase chain reaction (PCR) involves amplification of gene sequences with specific oligonucleotide primers. The convention is to use a temperature stable DNA polymerase from *Thermus aquaticus* (Taq) and to repeatedly cycle at low and high temperatures to replicate the DNA fragment, amplifying it to a level where it can be detected. Detection methods for the amplified probe include gel electrophoresis, immobilisation on membranes or direct probing, and visualisation by isotopic radiography, or fluorescent and chemiluminescent-labelling. The PCR technique offers the ability to specifically and rapidly detect pathogenic microorganisms from environmental samples without the isolation step on growth media. PCR may be particularly appropriate for detecting low numbers of pathogens in untreated and treated wastes, given the large background of competing species present. However, PCR detection data must be interpreted with caution because the technique detects not only vegetative or VNC microorganisms but also dead cells whose DNA has yet to degrade.[40] Consequently, in sewage sludge treated to the appropriate standards, e.g. using 'disinfection' with heat or UV irradiation, the dead microorganisms will continue to be detected by PCR; this is analogous to a obtaining a false positive result. This discrepancy has been demonstrated in a study to assess the survival of *E. coli* B and *S. typhimurium* Q in aerobic thermophilic composts using DNA gene probes.[41] The authors observed that the DNA targets in food waste compost and sludge compost were still detectable after many days at 60-70°C. Therefore, PCR may be a better diagnostic tool for the microbiological quality of untreated wastes than for those undergoing subsequent treatment.

Furthermore, care must be taken with PCR based assays because of the potential for inhibitory substances present in faeces and other wastes to produce false negative reactions, including iron and polyphenols.[42] Environmental matrices can be complex and undefined, making a generic clean-up procedure difficult. For sludge samples, Tsai *et al*.[43] recommended rapid freeze-thawing, followed by a phenol-chloroform extraction and clean-up through a Sephadex-200 spin column to obtain adequate recovery of target DNA. Reverse transcriptase PCR (RT-PCR) detection is being increasingly used to detect RNA enteroviruses in water, and Stewart and Abbaszadegan[44] advocated clean-up through a Sephadex G-100 column to remove humic acids, and through a Chelex-100 column to remove ionic inhibitors, for good detection of the viruses. Humic acids can also be removed during the extraction process by adding polyvinyl-polypyrrolidone; this procedure also increases the yields and quality of the extracted DNA.[45]

The PCR technique has been used to detect *E. coli* and *Shigella* spp. in water using gene probes for the *uidA* gene[46], as well as sewage and sludge.[43] The latter workers obtained a sensitivity of detection of 2.5-4 cells g^{-1} extracted sludge. The use of the *uidA* gene as a suitable marker was confirmed by Fricker and Fricker [47] who performed PCR on single colonies of 324 strains of coliforms and 117 of *E. coli* grown overnight on MacConkey agar. The basic PCR protocol was 25 cycles of 60 sec at 94°C, 60 sec at 60°C and 60 sec at 72°C. The primer set which was derived from the *uidA* gene correctly identified all of the *E. coli* strains tested; however, the sequence was also identified in 5 non-*E. coli* coliforms.

4.5 Immunollabelling methods

The immunolabelling methods fall into several broad categories, summarised as immunofluorescence assay (IFA), enzyme immunoassay (EIA: also commonly referred to as enzyme-linked immunosorbent assay; ELISA) and latex bead agglutination (LBA). IFA can be used for the direct specific **detection** of original samples, providing the assay is sufficiently sensitive for the concentration of target organism present, and after sample concentration (filtration or immunocapture) or amplification (pre-enrichment) steps. EIA (ELISA) and LBA are used more often for rapid post-broth and agar colony **screening** to confirm the identity of the target organism, augmented by conventional serotyping (see later).

IFA offers the ability to specifically detect pathogens *in situ*, particularly where sub-lethally damaged or VNC bacteria are suspected. Once the antibody has been produced, the method is quick and inexpensive. Recently, IFA has been combined with CTC-detected respiration to determine the identity and physiological status of *E. coli* O157 in water.[48] Fluorescently-labelled antibodies have been used to screen wastewater for *Salmonella*.[49] The method is rapid but requires that there is no cross-reactivity with other species, that the target epitope is expressed and conserved in the test environment and there are sufficient number of cells for observation by microscopy or cell cytometry (see later). The IFA technique has recently been used to show that *C. jejuni* survives in tap water or aquatic biofilms with a high species diversity for over 28 days at 4° or 30°C.[17] However, care must be taken with direct epifluorescence microscopy observation if there is a high background fluorescence due to autofluorescence of the sample or non-specific binding of the antibody to the autochthonous microflora. Under these circumstances, cell cytometry may provide greater discrimination of homogenised samples.

4.6 Emerging technologies – cell and laser scanning cytometry

Cell cytometry relies on a stream of liquid flowing as discrete microdroplets through a laser beam. Optical signals are detected whenever a particle, either unlabelled or fluorescently labelled, passes through at rates exceeding 10,000 per second. The types of information available include size, shape, labelled RNA, DNA and surface antigen content. The data are collected for comparison of parameters such as size versus fluorescent intensity. Incorporation of a fluorescently activated cell sorter (FACS) allows gates to be set of say size versus fluorescence and each particle which gives a positive signal within the gate can be deflected to a collector to provide a specific separation and quantification procedure. The specificity of fluorescently labelled antibodies has been exploited to detect and purify microorganisms such as *E. coli*, *L. pneumophila*, spores of *Bacillus anthracis* and oocysts of *Cryptosporidium parvum* by flow cytometry.[50, 51] The coupling of flow cytometry with the use of 16S or 18S rRNA fluorescent probes has been advocated to facilitate the quantification of specific microorganisms from environmental samples.[39, 51]

A system that addresses the need for rapid detection and identification of microorganisms from environmental samples has been developed by Chemunex (Maisons Alfort, France). The ChemScan RDI is based on direct fluorescent labelling of viable organisms trapped on a 25 mm diameter membrane, coupled with an ultra-sensitive laser scanning and counting system. The high level of sensitivity of the solid phase cytometer means that a single cell on a membrane can be detected. The use of fluorescently-labelled antibodies, enzyme substrates or nucleic acid probes provides the specificity for ChemScan to identify and enumerate target microorganisms without the need for enrichment. FITC-labelled *C. parvum* oocysts can be counted within 3 minutes before visual observation of the presumptive positives by epifluorescence microscopy.[52] Viability can be assessed by incubating with fluorochrome esters which fluoresce when the substrate is actively taken up by viable cells and intracellular esterases release the fluorochrome. The technology is now ready for application to untreated and treated wastes, provided good fluorescent antibody and oligonucleotide reagents are available, and trapped non-target cells do not interfere. The latter may be unlikely when looking for low numbers of a pathogen against a high background in sewage sludge, unless IMS or selective enrichment is undertaken first. Reynolds *et al.*[52] observed that one advantage of isolating target organisms on a membrane was that interfering substances such as clay particles could be washed away before incubation with antibody reagents.

4.7 Pre-enrichment of low numbers/stressed cells

The numbers of pathogens in wastewater and biosolids varies according to treatment and age. For example, the numbers of *Salmonella* and *Campylobacter* spp. in raw sewage range from 7 to 8000 per 100 ml, and occasionally higher.[53, 54] In activated sludge and digested sludge, *Salmonella* and *Campylobacter* cell numbers range from none detected to a most probable number detected of 400 cells per gram dry weight of solids.[54, 55] As a consequence of these low numbers, some form of enrichment, perhaps preceded by a sample concentration step, is required. Conventional isolation protocols involve 3 steps:

1. Pre-enrichment - incubation for 18-24 hours in a nutrient broth such as 1% buffered peptone. This allows the multiplication of organisms towards detectable levels (dependent on the sensitivity of the subsequent assay procedure)
2. Selective enrichment - incubation for 24-48 hours in a combination of selective broths. The selective broth allows the multiplication of the organism of interest and has the effect of reducing the level of background organisms whilst increasing the level of the target organism.
3. Isolation from selective broth by the use of selective agar.

However, steps 2 and 3 may still impart a severe stress on the target organism which may not grow if it is sublethally damaged due to exposure to environmental factors such as nutrient starvation or UV irradiation, oxygen, pH or disinfectant stress. A typical procedure for pre-enriching *E. coli* O157, *Shigella* and *Salmonella* spp. involves homogenisation of 25g of sample in 225 ml of 1% buffered peptone water (BPW) or tryptone soya broth (TSB).[56] This is incubated at 37°C for 16-18 hours, preferably with agitation (150 rpm) before subculture to selective media for further enrichment and/or isolation. A summary of pre-enrichment methods for several pathogens which may be found in sludge and animal wastes is presented in Table 3.

Campylobacter spp. are microaerophilic and it is an important consideration to pre-enrich and subsequently isolate in a low oxygen environment. Suitable enrichment broths contain nutrient broth (Oxoid No. 2), lysed blood and FBP supplement (ferrous sulphate, sodium metabisulphite and sodium pyruvate, each at 0.025% (w/v)) to improve aerotolerance. A mixture of antibiotics, such as the trimethoprim, rifampicin, polymixin and amphotericin, is also required to prevent overgrowth of slow growing *C. jejuni* by competing organisms; these antibiotics are included in Preston broth.[57] *Campylobacter* spp. were successfully enriched from untreated and treated sewage in Preston broth incubated at 43°C, followed by isolation on Preston agar under microaerophilic conditions.[58] Exeter broth also includes cefoperazone and gives superior isolation rates. Sensitivity to some of the antibiotics demonstrated by sub-lethally damaged campylobacters can be overcome by incubating the broths at 37°C rather than 42°C.[59] Similarly, *L. monocytogenes* can be recovered from raw and treated sewage in higher numbers if first resuscitated in FDA Listeria broth (Oxoid) in a microaerophilic environment before being enumerated on a chromagenic agar such as BCM Listeria medium (Biosynth, Switzerland).[60]

4.8 Specific immunomagnetic capture

Latterly, researchers have attempted to overcome the problems of interference from the background matrix, lack of sensitivity of detection and the long process of enrichment by using selective separation with antibodies liganded to magnetic particles. There are 2 principal companies involved in immunomagnetic separation (IMS) of pathogens, Dynal and IDG, and both supply kits which require pre-enrichment of samples in broth culture before capture on superparamagnetic polystyrene beads linked to antibodies. The beads are

Table 3 *Methods to resuscitate, enumerate and identify bacterial pathogens*

Pathogen	Resuscitate/ enrich (37°C)	Discriminate	Confirm
VTEC	BPW/TSA ± Novobiocin	SMAC Chromagars 41°C	Ab serotype Ab toxin Phage type RFLP
Salmonella	BPW/TSA Rapp-Vass	xyl lys desoxy, mod brill green 41°C	Ab serotype Phage type
Shigella	BPW/TSA Hajna-GN	Hektoen, Desoxy citrate	Ab serotype Biochem profile
Campylobacter	Exeter broth, low O_2	Preston agar 42°C, low O_2	Ab serotype Biochem profile
Listeria	FDA broth, low O_2 better	BCM agar, low O_2 better	Ab serotype Biochem profile

designed to replace the use of selective enrichment broths, and produce about the same degree of enrichment within 30 minutes as opposed to 24 hours. At the appropriate time, powerful magnets draw the beads to one side of the incubation tube allowing the supernatant containing unwanted material to be aspirated. The beads can then be washed before further analysis of the captured pathogens by PCR, ELISA, staining and microscopy, or culture.

The technology involved in coupling monoclonal or polyclonal antibodies to magnetic beads for the IMS techniques is well established and has been used to detect salmonellae in food[61], and biotoxoids and bacterial spores.[62] More recently it has been advocated for the detection of low numbers of *C. parvum* and *Giardia lamblia (intestinalis)* in potable water, post-filtration, and forms the basis of US EPA Methods 1622 and 1623. IMS has also found favour for the selective detection of *E. coli* O157 in food and water[63] and faeces[64]; the detection limit was 1-2 cfu g^{-1} sample. Cubbon *et al*.[65] found that IMS detection of O157 in faecal samples was more sensitive than culture and compared well with PCR. The main problem when using the IMS technique is the number of sorbitol non-fermenters other than *E. coli* O157 that adhere non-specifically to the magnetic beads. Recovery of the pathogen from enrichment broth is enhanced by using antibody-coated magnetic beads and non-specific binding of other organisms is reduced by washing beads with phosphate buffered saline containing 0.002-0.05% Tween 80.[3]

4.9 Selective enrichment

Non-selective media such as nutrient broth are widely presumed not to stress sub-lethally damaged bacteria, thereby enhancing the efficiency of recovery. However, most complex biosolids and wastes contain a diverse, complex microbial flora which can overgrow the target microorganism, particularly if it grows slowly or is present in low numbers. Accordingly, modified buffered peptone water (MBPW) is recommended for the enrichment of *E. coli* O157 from complex biosolids such as cattle faeces[66] (Table 3). The 1% peptone medium is supplemented with vancomycin
(8 mg l^{-1}), cefixime (0.05 mg l^{-1}) and cefsulodin (10 mg l^{-1}). Recovery of *E. coli* O157 from enrichment broth is improved by using immunomagnetic separation with antibody-coated magnetic beads.

Enrichment media for *Salmonella* include Rappaport-Vassiliadis broth (RV) containing malachite green and Selenite cystine broth (SC).[67] These form part of the British Standards Institute method for analysis of food and animal feeding stuffs.[68] Samples are incubated in RV at 42°C and in SC at 37°C for 18-24 hours before subculture to selective agar. The incubation temperature can significantly influence the recovery efficiency of selective enrichment media. For many years it has been known that *Salmonella* recoveries can be enhanced by using incubation temperatures of between 40 and 43°C.[69] The use of elevated temperatures must be matched with the enrichment system used. For example, elevated temperatures do not enhance recovery in selenite brilliant green or tetrathionate-based broths, but do work well with RV and have been incorporated into the secondary enrichment method for detecting *Salmonella* spp. in composted sewage sludge.[70] Although rapid end-point tests enable *Salmonella* to be detected in half an hour (see later), even the most sensitive of them requires at least 10^4 cells ml^{-1} of broth. Most naturally contaminated foodstuffs or environmental samples contain far fewer stressed cells ml^{-1}, making the initial enrichment phase essential.

4.10 Impedance detection

Impedance technology is a rapid, automated qualitative technique which measures in a medium the conductance change induced by bacterial metabolism.[71] The detection time is a function of both initial microorganism concentration and growth kinetics in a given medium. Specificity is incorporated into the technique by including either selective agents into the incubation broth and/or specific substrates. Thus, Easter and Gibson [72] described an impedance technique in which changes in electrical conductance due to reduction by salmonellae of trimethylamine-N-oxide were monitored. By contrast, Bullock and Frosham [73] pre-enriched salmonellas from contaminated confectionery in skimmed milk before 24 hour impediometry in lysine-iron-cystine-neutral red broth in a Bactometer 123 system (Bactomatic Ltd., Henley). The authors found that the inclusion of novobiocin (0.15 µg per well) eliminated false positive results due to *Citrobacter freundii* or *Enterobacter cloacae*. Pridmore and Silley [74] used the Rapid Automated Bacterial Impedance Technique (RABIT, Don Whitley, Shipley) to detect total coliforms, thermotolerant coliforms and enterococci in domestic sewage and 70% industrial sewage from 2 wastewater treatment works. The coliforms were detected in Whitley MacConkey broth at 37°C and 44°C using the direct impedance technique. The majority of faecal coliform results were obtained within 7 hours (10^3 cfu ml^{-1}) compared to 24 hours using membrane filtration on MLSB, and without the need for serial dilution of samples and manual reading of plates.

The indirect impedance technique allows the use of components inappropriate in the direct method on account of high basal conductance. This method is based on the detection of carbon dioxide released by microorganisms into the culture medium, and which is absorbed in an alkaline solution in contact with the electrodes of the tubes. Recently, Blivet *et al.* [75] proposed a new medium named KIMAN (Whitley Impedance Broth basal medium supplemented with 3 selective components: potassium iodide, malachite green and novobiocin,). This medium supported the growth of *Salmonella* serotypes, while inhibiting non-salmonella strains in pure culture, and was appropriate for the indirect impedance technique.

4.11 Immunoscreening

A large proportion of immunolabelling is used as culture confirmation of species identity following enrichment and agar culture, using enzyme immunoassay (EIA) (also called enzyme linked immunosorbent assay; ELISA) or latex bead agglutination (LBA) augmented with serotyping and phage typing. However, these techniques can also be used without the agar culture step to speed up the recovery/detection time. A typical format for EIA or ELISA involves the coating of rabbit polyclonal or mouse monoclonal antibody to the wells of microtitre plates followed by introduction of the test sample. Protein or polysaccharide (lipopolysaccharide) antigens present in the sample are bound immunologically by the antibody. After washing to remove unbound material, enzyme-conjugated affinity-purified antibody specific to the target antigen is added. Following a second washing step to remove unbound enzyme-conjugated antibody, enzyme substrate is added and the incubation proceeds until stopped e.g. by addition of acid or alkali which also helps develop the product colour. Typical enzymes used include alkaline phosphatase (with p-nitrophenyl phosphate as substrate) and horseradish peroxidase (with 3,3',5,5'-tetramethylbenzidine (TMB) as substrate and hydrogen peroxide) Most EIA kits have a sensitivity of approximately 10^6 organisms ml^{-1}, and therefore usually require a concentration step (filtration, IMS and/or pre-enrichment). Pre-enrichment in appropriate medium may be obligatory if there are concerns that epitope expression, e.g. flagellar antigen, is affected by the environment.

Many rapid manual immunoscreening assays are commercially available to detect *E. coli* O157 and *Salmonella* spp. These tests are performed on heat-killed culture broth after 24 hours for *E. coli* O157 and 40-48 hours for *Salmonella* spp. Microtitre well-based ELISAs (e.g. Microgen Salmonella ELISA) and dipstick ELISAs (e.g. Lumac Salmonella Path-Stick) are also commercially available for these pathogens and are recommended for wastewater treatment. Organon Teknika have gone a step further by introducing an ELISA for *E. coli* O157 incorporating immuno-capture beads (EHEC-Tek). Recently, the manual ELISA tests have been adapted for use in automated instruments and greatly increase the capacity of a laboratory to perform up to 100,000 tests per annum on one instrument. ELISA has been used to detect the presence of enterotoxigenic *E. coli* in water[76] and *S. enteritidis* in raw sewage, sludge and wastewater (Brigmon *et al.*, 1992).[77] ELISA technology is maturing rapidly and can be included in 96-well plates for automated reading and software manipulation. However, a major disadvantage of the technique is the lack of sensitivity. A minimum of 10^5 *S. enteritidis* ml^{-1} are required to generate a clear signal against the background.

LBA provides the least technically demanding method and, as the name suggests, relies on the agglutination of microscopic latex beads which are liganded with a specific polyclonal or monoclonal antibody to an epitope expressed by the microorganism. The preparations become cloudy or clump, which can be seen against a dark background. Sensitivity of detection varies from 10^2 - 10^6 cells ml^{-1}, depending on the avidity of the antigen-antibody reaction. For example, *E. coli* O157 can be detected with latex bead agglutination using antibodies raised against the lipopolysaccharide O157 antigen, reversed passive latex agglutination and passive haemagglutination. Isolates can also be serotyped with antisera and phage typed, as demonstrated successfully by Rahn *et al*.[78] in a detailed study of *E. coli* O157:H7 persistence in human and animal faeces.

4.12 Selective agar media

Various selective agar media have been developed for the range of bacterial pathogens in sludge and animal wastes (Table 3). Most strains of VT *E. coli* O157 do not ferment sorbitol, and sorbitol-MacConkey (SMAC) medium was devised to facilitate their isolation.[79] Almost inevitably, however, some other serogroups of *E. coli* are also sorbitol non-fermenters (NSF), as are members of several other genera found frequently in the faeces of man and animals, particularly *Proteus* and *Aeromonas* spp. Media have been developed to decrease the number of NSF that need to be screened during the attempted isolation and quantification of *E. coli* O157. These include cefixime/rhamnose/sorbitol MacConkey (CR-SMAC) medium[80], in which cefixime inhibits *Proteus* spp. at a concentration not inhibitory to *E. coli* and rhamnose is fermented by most NSF *E. coli* strains of serogroups other than O157, and β-glucuronidase medium in which colonies of VT *E. coli* can be detected by their lack of fluorescence.[63]

Tellurite has been used for most of the century for the isolation of pathogens such as *Shigella* spp.[81] It has subsequently been demonstrated to be selective for VT *E. coli* O157 and *S. sonnei* but partially or completely inhibited many other strains of *E. coli* and most strains of NSF.[82] These workers found that inclusion of potassium tellurite in SMAC markedly increased the rate of isolation of VT *E. coli* O157 from cattle rectal swabs. The tellurite concentration was critical and incorporation of 2.5 mg l^{-1} into SMAC containing 0.05 mg l^{-1} cefixime (CT-SMAC) gave optimal recoveries of VT *E. coli* O157, compared to SMAC or CR-SMAC, and inhibited NSF such as *Aeromonas* and *Providencia* spp. The inclusion of cefixime inhibited the growth of *S. sonnei*. Not surprisingly, therefore, CT-SMAC has become the selective medium of choice for the recovery of VT *E. coli* O157 from complex biosolids.

Unlike most coliforms, those *Salmonella* pathogenic to man (including *S. enteritidis* PT4 *and S. typhimurium* DT 104) do not ferment lactose and are β-galactosidase negative. Media selective for *Salmonella* usually include selenium salts, brilliant green or malachite green to inhibit other genera. Typical examples are modified brilliant green agar (BGA), xylose lysine desoxycholate agar (XLD), desoxycholate citrate agar (DCA), salmonella-shigella agar (SS), brilliant green MacConkey agar (BGM), bismuth sulphite agar (BSA) and mannitol lysine crystal violet brilliant green (MLCB).

4.13 Chromagenic media

Great advances have been made in the formulation of agar media containing chromagens which react with spceific enzymes of target microorganisms, conveying specificity of detection, ease of use and reduced man power. For example, CHROMagar® (M-Tech Diagnostics, Warrington or Becton-Dickinson, USA) gives clearly differentiated violet colonies with *E. coli* O157, blue with other *E. coli*, red with other coliforms and colourless with non-coliform Gram negative bacteria. Rainbow Agar O157 (BIOLOG, USA) detects toxin-producing serotypes of *E. coli*, particularly O157:H7. The agar includes chromogenic substrates for β-galactosidase and β-glucuronidase. Most *E. coli* are β-glucuronidase-positive and produce red/magenta colonies on the agar. Strain O157:H7 is typically β-glucuronidase-negative and forms characteristic charcoal grey/black colonies. Other toxigenic strains such as O111 overproduce β-galactosidase relative to β-glucuronidase and are typically coloured purple/violet/blue. Addition of novobiocin at 10 mg l^{-1} slightly increases the selectivity of the medium by further inhibiting the background growth of other species in faeces such as enterococci. Increasing the novobiocin concentration to 100 mg l^{-1} renders the medium highly selective for strain O157:H7 since most other *E. coli* and the faecal background bacteria are inhibited for at least 24 hours.

Rambach agar[83] utilises a novel phenotypic characteristic, the formation of acid from propylene glycol, and hydrolysis of X-gal to differentiate *Salmonella* spp. from other members of the *Enterobacteriaceae*. While *Salmonella* colonies are bright red, coliforms appear blue, green, violet or colourless. The sensitivity of this agar to detect salmonellae varied from 69%[84] to 91%[85] but with a specificity of 98-100%. Accordingly, Manafi and Sommer [85] recommended that Rambach agar was not appropriate as a primary plating medium when screening for non-typhi salmonellae, due to variable sensitivity, but its high specificity made it suitable for plating after enrichment. Pignato *et al.*[86] used Salmosyst broth (Merck) as a combined pre-enrichment/selective enrichment broth and Rambach agar for isolation. They found that in artificially contaminated ground beef *S. enteritidis* was detected at a concentration of 10 cfu per 25 g.

BIOLOG's Rainbow Agar Salmonella is now available for the successful isolation and differentiation of the widest range of *Salmonella* spp., including *S. typhi*, in pure and mixed cultures The agar medium exploits the well-established characteristic of H$_2$S production to detect *Salmonella* but the chemical reaction has been intensified so that even weakly H$_2$S-producing strains develop clearly differentiated colonies.

5 FUTURE DEVELOPMENTS

Clearly, great progress is now being made in the adaptation of existing methods and the development of new sensitive and quantitative detection technologies for water and faecal wastes. Some areas which are now being considered for further development are listed in Table 4. These involve reviewing how samples should be spiked for validation studies, for example should bacterial inocula for spiking studies be grown anaerobically to pre-adapt them to the more physiologically relevant environment of faeces? Is it possible to prepare natural or spiked samples to ensure complete release and capture of the pathogen from a complex matrix while still maintaining its viability for culture recovery? This again requires that more knowledge of microbial physiology is required to aid the formulation of suitable release buffers and the development of suitable resuscitation pre-culture and agar media. IMS shows great promise but is obviously dependent on expression of the appropriate cell surface epitope for efficient capture; is epitope expression affected by the pre-enrichment conditions? Similar concerns apply to EIA and IMS-EIA detection technologies. Molecular detection is claimed by many to be the technology platform for the future, but PCR still suffers from the deleterious effects of inhibitors such as iron or humic acids being present in the sample. Moreover, PCR and 16s/18s rRNA FISH do not necessarily detect viable cells. Monitoring mRNA content using RT-PCR has been suggested to show that a cell is still viable due to the rapid turnover over the molecule, but some experiments have shown that a cell can be killed without the mRNA degrading rapidly.[87] Similarly, rRNA content can be stable in dead cells for quite some time, so that a high content does not necessarily mean that the cell is still viable and has survived any storage or treatment procedure. Even the well established enzyme assays with chromogenic and fluorogenic –labelled substrates do not define the viability of cell; after all a dead cell is still a bag of enzymes, some of which may have retained activity. For this reason, even some of the so-called viability assays involving cellular respiration or dye uptake/exclusion (as in the case of the DAPI/PI assay for *Cryptosporidium* and *Giardia* oocysts/cysts) should be viewed with caution since they are really not viability assays but vitality assays. Nevertheless, science has made great progress in the past 20 years in the detection of pathogens in water and food, and this knowledge will provide appropriate guidance and enable the formulation of suitable legislation to reassure the consumer and protect the public health.

Table 4 *Ongoing studies for biosolids*

Method	Procedure
Sample spiking	grow anaerobically (physiologically relevant)
Sample preparation	clean pathogen release; maintain viability
Media formulation	resuscitation of stressed cells
IMS	pre-enrichment (but antibody dependent)
Antibody detection	EIA, IMS-EIA
Molecular detection	PCR (but inhibitors?); 16s/18s rRNA FISH
	specific, sensitive (but detect "viable" cells?; mRNA
Enzyme activity	β-galactosidase, β-glucuronidase; C8 esterase
Vitality activity	respiratory (CTC), ester uptake (+ esterase)
(cell cytometry)	dye exclusion (DAPI/PI)?

References

1   P. J. Packer, C. W. Mackerness, M. Riches, and C. W. Keevil, *Lett. Appl. Microbiol.*, 1995, **20**, 303.

2   PHLS/SCA, *J. Hyg.*, 1980, **85**, 1181.

3   Anon, 'The Microbiology of Water 1994. Part 1- Drinking Water. Report on Public Health and Medical Subjects No.71.', HMSO, London, 1994.

4   APHA, 'Standard Methods for the Examination of Water and Waste Water', American Public Heath Association, Washington DC, 1989.

5   M. W. LeChevallier, S. C. Cameron, and G. A. McFeters, *Appl. Environ. Microbiol.*, 1983, **45**, 484.

6   A. Mates and M. Shaffer, *J. Appl. Bacteriol.*, 1989, **67**, 343.

7   D. P. Sartory and L. Howard, *Lett. Appl. Microbiol.*, 1992, **15**, 273.

8   K. S. Walter, E. J. Fricker, and C. R. Fricker, *Lett. Appl. Microbiol.*, 1994, **19**, 47.

9   N. Humphrey, 'A survey of *E. coli* in UK sludges', UK Water Industry Research Ltd., London, 1999.

10   F. A. Nicholson, M. L. Hutchison, K. A. Smith, C. W. Keevil, B. J. Chambers, and A. Moore, 'A study of farm manure applications to agricultural land and an assessment of the risks of pathogen transfer into the food chain', HMSO: MAFF Publications, London, 2000.

11   Day, in 'Fields of filth', New Scientist, 1998, 4th February, p. 8.

12   Anon, 'The Advisory Committee on the Microbiological Safety of Food. Report on verocytotoxin-producing *Escherichia coli.*', HMSO, London, 1995.

13   A. Maule, in 'Coliforms and *E. coli*: Problem or Solution', Royal Society of Chemistry, Cambridge, 1997, p. 61.

14   C. W. Keevil, J. T. Walker, A. Maule, and B. W. James, in 'Verocytotoxigenic *E. coli* in Europe: Survival and Growth', ed. P. G. G. Duffy, J. Coia, W. Wasteson and D. McDowell, Teagasc, Dublin, 1999, p. 42.

15   J. Rogers and C. W. Keevil, in 'Protozoal Parasites in Water', ed. K. C. Thompson and C. Fricker, Royal Society of Chemistry, London, 1995, p. 209.

16   C. W. Keevil, Rogers, J. and Walker, J.T., *Microbiology Europe*, 1995, **3**, 10.

17   C. M. Buswell, Y. M. Herlihy, L. M. Lawrence, J. T. McGuiggan, P. D. Marsh, C. W. Keevil, and S. A. Leach, *Appl. Environ. Microbiol.*, 1998, **64**, 733.

18   D. Roberts, W. Hooper, and M. Greenwood, 'Practical Food Microbiology', PHLS, London, 1995.

19   P. Bracegirdle, A. A. West, M. S. Lever, R. B. Fitzgeorge, and A. Baskerville, *Epidemiol. Infect.*, 1994, **112**, 69.

20   C. V. Broome, C. A. Cieseilski, M. J. Linnan, and A. W. Hightower, 73rd Annual Meeting of the IAMFES, 1986.

21   R. G. D'Antonio, R. E. Winn, J. P. Taylor, T. L. Gustafson, W. L. Current, M. M. Rhodes, G. W. Gary, Jr., and R. A. Zajac, *Ann. Intern. Med.*, 1985, **103**, 886.

22   W. R. Mackenzie, N. J. Hoxie, M. E. Proctor, M. S. Gradus, K. A. Blair, D. E. Peterson, J. J. Kazmierczak, D. G. Addiss, K. R. Fox, J. B. Rose, and J. P. Davis, *N. Engl. J. Med.*, 1994, **331**, 161.

23   M. M. Peng, L. Xiao, A. R. Freeman, M. J. Arrowood, A. A. Escalante, A. C. Weltman, C. S. Ong, W. R. Mac Kenzie, A. A. Lal, and C. B. Beard, *Emerg. Infect. Dis.*, 1997, **3**, 567.

24   R. D. Adam, *Microbiol. Rev.*, 1991, **55**, 706.

25   M. Q. Deng and D. O. Cliver, *Parasito.l Res.*, 1999, **85**, 733.

26   B. Mignotte, A. Maul, and L. Schwartzbrod, *J. Virol. Methods*, 1999, **78**, 71.

27   Anon, *Fed. Regist.*, 1993, **58**, 9387.

28   C. W. Keevil, 'Review and status of current methods for the detection of verocytotoxic *Escherichia coli*, *Salmonella enteritidis* PT4 and *Salmonella typhimurium DT104*, *Shigella sonnei* and *Campylobacter jejuni* in treated and untreated biological effluents.', UK Water Industry Research Ltd, London, 1998.

29   K. Wadhia, A. Colley, and K. C. Thompson, in 'Toxic Impacts of Wastes on the Aquatic Environment', ed. K. C. Thompson, Royal Society of Chemistry, Cambridge, 1996.

30   Anon, 'Determination of the inhibitory effects of chemicals, industrial and other waste waters on the respiration of activated sludge using an oxystatic respirometer, Standing Committee of Analysts 1998.', Environment Agency, London, 1998.

31   R. R. Zimmerman, R. Iturriagu, and J. Becker-Birck, *Appl. Environ. Microbiol.*, 1978, **36**, 926.

32   G. Rodriguez, G. D. Phipps, K. Ishiguro, and H. F. Ridgway, *Appl. Environ. Microbiol.*, 1992, **58**, 1801.

33   A. N. Sharp and A. K. Jackson, *Appl. Microbiol.*, 1972, **24**, 175.

34   A. N. Sharpe, E. M. Hearn, and J. Kovacs-Nolan, *J. Food Prot.*, 2000, **63**, 126.

35   J. T. Staley and A. Konopka, *Ann. Rev. Microbiol.*, 1985, **39**, 321.

36   D. B. Roszak and R. R. Colwell, *Microbiol. Rev.*, 1987, **51**, 365.

37   G. A. McFeters, S. C. Broadaway, B. H. Pyle, M. Pickett, and Y. Egozy, *Water Sci. Technol.*, 1995, **31**, 259.

38   D. J. Reasoner and E. E. Geldreich, *Appl. Environ. Microbiol.*, 1985, **49**, 1.

39   R. I. Amann, W. Ludwig, and K. H. Schleifer, *Microbiol. Rev.*, 1995, **59**, 143.

40   K. L. Josephson, C. P. Gerba, and I. L. Pepper, *App.l Environ. Microbiol.*, 1993, **59**, 3513.

41   M. L. Droffner and W. F. Brinton, *Zentralbl. Hyg. Umweltmed.*, 1995, **197**, 387.

42   Y.-L. Tsai and B. H. Olson, *Appl. Environ. Microbiol.*, 1992, **58**, 2292.

43   Y. L. Tsai, C. J. Palmer, and L. R. Sangermano, *Appl. Environ. Microbiol.*, 1993, **59**, 353.

44   P. W. Stewart and M. Abbaszadegan, Abstracts of the 95th General Meeting of the American Society for Microbiology 1995, abstr. Q. 469, p.394.

45   R. J. Steffan, J. Goksoyr, K. B. Asim, and R. M. Atlas, *Appl. Environ. Microbiol.*, 1988, **54**, 2908.

46   A. K. Bej, S. C. McCarty, and R. M. Atlas, *Appl. Environ. Microbiol.*, 1991, **57**, 2429.

47   E. J. Fricker and C. R. Fricker, *Lett. Appl. Microbiol.*, 1994, **19**, 44.

48   B. H. Pyle, S. C. Broadaway, and G. A. McFeters, *Appl. Environ. Microbiol.*, 1995, **61**, 2614.

49   C. J. Desmonts, J. Minet, R. R. Colwell, and M. Cormier, *Appl. Environ. Microbiol.*, 1990, **56**, 1448.

50   A. P. Phillips and K. L. Martin, *J. Immunol. Meth.*, 1988, **5**, 205.

51   G. Vesey, N. Ashbolt, E. J. Fricker, D. Deere, K. L. Williams, D. A. Veal, and M. Dorsch, *J Appl. Microbiol.*, 1998, **85**, 429.

52 D. T. Reynolds, R. B. Slade, N. J. Sykes, A. Jonas, and C. R. Fricker, *J. Appl. Microbiol.*, 1999, **87**, 804.

53 R. G. D. Feachem, D. J. Bradley, H. Garelick, and D. D. Mara, 'Sanitation and Disease: Health Aspects of Excreta and Wastewater Management.', John Wiley & Sons, New York, 1983.

54 K. Jones, M. Betaieb, and D. R. Telford, *J. Appl. Bacteriol.*, 1990, **69**, 185.

55 W. A. Yanko, A. S. Walker, J. L. Jackson, L. L. Loabo, and A. L. Garcia, *Water Environ. Res.*, 1995, **67**, 364.

56 W. Edel and E. H. Kampelmacher, *Bull. World Health Org.*, 1973, **48**, 167.

57 F. J. Bolton and L. Robertson, *J. Clin. Pathol.*, 1982, **35**, 462.

58 S. M. Arimi, C. R. Fricker, and R. W. Park, *Epidemiol. Infect.*, 1988, **101**, 279.

59 T. J. Humphrey and I. Muscat, *Lett. Appl. Microbiol.*, 1989, **9**, 137.

60 K. J. Bown and C. W. Keevil, 'Methods for the detection of pathogens in biosolids. Report SL/06 to UK Water Industry Research Ltd, London', 2000.

61 J. M. Luk and A. A. Lindberg, *J. Immunol. Meth.*, 1991, **137**, 1.

62 D. L. Gatto-Menking, H. Yu, J. G. Bruno, M. T. Goode, M. Miller, and A. W. Zulich, *Biosensors and Bioelectronics*, 1995, **10**, 501.

63 A. J. G. Okrend, B. R. Rose, and C. P. Lattuada, *J. Food Protect.*, 1992, **55**, 214.

64 P. A. Chapman and C. A. Siddons, *J. Med. Microbiol.*, 1996, **44**, 267.

65 M. D. Cubbon, J. E. Coia, M. F. Hanson, and F. M. Thompson-Carter, *J. Med. Microbiol.*, 1996, **44**, 219.

66 P. A. Chapman, C. A. Siddons, D. J. Wright, P. Norman, J. Fox, and E. Crick, *Epidemiol. Infect.*, 1993, **111**, 439.

67 P. Vassiliadis, *J. Appl.Bacteriol.*, 1983, **54**, 69.

68 Anon, 'BS 5763. Microbiological examination of food and animal feeding stuffs. Part 4. Detection of *Salmonella*', British Standards Institute, London, 1993.

69 R. W. S. Harvey and T. H. Price, *J. Appl. Bacteriol.*, 1979, **46**, 27.

70 D. Hussong, W. D. Burge, and N. K. Enkiri, *Appl. Environ. Microbiol.*, 1985, **50**, 887.

71 P. Silley and S. Forsythe, *J. Appl. Bacteriol.*, 1996, **80**, 233.

72 M. C. Easter and D. M. Gibson, *J. Hyg.*, 1985, **94**, 245.

73 R. D. Bullock and D. Frodsham, *J Appl Bacteriol*, 1989, **66**, 385.

74 A. M. Pridmore and P. Silley, 6th Microbiological Methods Forum, Camden and Chorleywood Food Research Association, Chipping Campden, March 1998.

75 D. Blivet, G. Salvat, F. Humbert, and P. Colin, *J. Appl. Microbiol.*, 1998, **84**, 399.

76 Z. Tamanai Shacoori, A. Jolivet Gougeon, M. Pommepuy, M. Cormier, and R. R. Colwell, *Can. J. Microbiol.*, 1994, **40**, 243.

77 R. L. Brigmon, S. G. Zam, G. Bitton, and S. R. Farrah, *J. Immunol. Methods*, 1992, **152**, 135.

78 K. Rahn, S. A. Renwick, R. P. Johnson, J. B. Wilson, R. C. Clarke, D. Alves, S. McEwen, H. Lior, and J. Spika, *Epidemiol. Infect.*, 1997, **119**, 251.

79 S. B. March and S. Ratnam, *J. Clin. Microbiol.*, 1986, **23**, 869.

80 P. A. Chapman, C. A. Siddons, P. M. Zadik, and L. Jewes, *J. Med. Microbiol.*, 1991, **35**, 107.

81 M. Mujibur Rahaman, M. Golam Morshed, K. M. Sultanul Aziz, and M. M. H. Munshi, *Lancet*, 1986, **1**, 271.

82 P. M. Zadik, P. A. Chapman, and C. A. Siddons, *J. Med. Microbiol.*, 1993, **39**, 155.

83 A. Rambach, *Appl. Environ. Microbiol.*, 1990, **301-303**, 301.

84 H. Dusch and M. Altwegg, *J. Clin. Microbiol.*, 1993, **31**, 410.

85 M. Manafi and R. Sommer, *Lett. Appl. Microbiol.*, 1992, **14**, 163.

86 S. Pignato, A. M. Marino, M. C. Emanuele, V. Iannotta, S. Caracappa, and G. Giammanco, *App.l Environ. Microbiol.*, 1995, **61**, 1996.

87 G. E. Sheridan, C. I. Masters, J. A. Shallcross, and B. M. MacKey, *Appl. Environ. Microbiol.*, 1998, **64**, 1313.

SOLID PHASE CYTOMETRY FOR RAPID DETECTION OF TOTAL COLIFORMS AND *E. coli,* INCLUDING O157:H7, IN WATER AND FOOD

B.H. Pyle, and G.A. McFeters

Microbiology Department
Montana State University
Bozeman, Montana 59717
United States of America

S.O. Van Poucke and H.J. Nelis

Laboratory for Pharmaceutical Microbiology
University of Ghent
Ghent
Belgium

I INTRODUCTION

Culture methods for microbiological monitoring of indicator and pathogenic bacteria in food and water have well-known deficiencies, including the failure to detect a large proportion of the target organisms in a sample and the time taken to obtain confirmed results. The numbers of bacteria may be underestimated due to sublethal environmental injury and other factors which reduce culturability[1]. Most existing standard methods for analysis of water and foods require enrichment followed by confirmation or identification of the contaminating bacteria.

Detection of *Escherichia coli* and total coliforms (TC) in drinking water is performed by using either conventional lactose fermentation-based tests (e.g. m FC agar, m Endo agar LES), enzymatic presence-absence tests (e.g. Colilert, Fluorocult) or enzymatic membrane filtration tests (e.g. m Coliblue 24, Chromocult Coliform agar). Although the latter two afford a shorter analysis time (18-24 h) with regard to the conventional 24-48 h tests, they remain too time-consuming in case of emergency situations like water supply breakdowns or accidental contamination. Analysis of foods is even more problematic, because the matrix contains many components including fibres and particles which may clog membrane filters and mask bacterial growth. These factors contribute to the lack of sensitivity and reliability of routine culture methods.

Direct detection of bacteria in food and water avoids the need for subculture and enrichment which are time consuming and tend to underestimate because of injury and other factors. Staining bacteria with fluorochromes facilitates rapid, direct detection of the organisms of interest. Using epifluorescence microscopy, it is possible to obtain more timely estimates of contaminants. However, this method is not very sensitive.

Recently developed staining techniques may allow an estimate of either the numbers of particular bacteria, or their activity. Both assessment of metabolic activity and identification of the organisms are essential to be sure of what organisms are present and their physiological status. In addition, more sensitive detection may be achieved by the use of techniques such as solid phase laser cytometry (ChemScan RDI, Chemunex, Maisons-Alfort, France). Techniques which use such an instrument for the detection of total coliforms and *E. coli* in water, verocytotoxic *E. coli* in water

and foods, and total metabolically active bacteria have been developed using the laser
scanning system described below.

II Solid Phase Cytometry

In solid phase cytometry, fluorescently labelled microorganisms immobilized on a
membrane filter are detected by a laser scanning system. The ChemScan RDI is such a
laser-scanning device that visualizes and enumerates fluorescent cells retained on a
25-mm membrane filter within 3 min[2]. The lower level of detection on an entire filter
is a single cell. Following scanning, the results obtained can be validated by using an
epifluorescence microscope directly connected to the ChemScan system and equipped
with a motorized stage that is driven by the system software. This enables the operator
to visually confirm the detected microorganisms or particles present on the membrane
filter in order to confirm the software-based discrimination process. The whole
validation process can be performed within seconds to a few minutes, the actual time
depending on the numbers of microorganisms present.

III Total Coliforms and *E. coli* in water

By using the ChemScan system in conjuction with two-stage enzymatic detection
methodology, designed to obtain optimal enzyme induction and cell labelling in two
separate steps[3,4], total coliforms and *E. coli* in water can be enumerated within 3.5 h.
The procedure involves enzyme induction (3 h) in the target cells trapped on the
polyester membrane filter, followed by cell labelling (30 min) using fluorescein-di-ß-
D-glucuronide (*E. coli* detection) or fluorescein-di-ß-D-galactopyranoside (TC
detection) and laser scanning (3 min). A prevalidation study with 50 naturally
contaminated well and surface water samples has indicated good agreement (>93%)
with reference methods, including m FC agar (*E. coli*), m Endo agar LES (TC) and
Chromocult Coliform agar (*E. coli* and TC). The false-negative error was only 1.9%.
Application of the protocols to uncontaminated tap and well water samples has not
revealed any false-positive results yet. These methods are promising for rapid
detection of *E. coli* and TC in drinking water.

III *E. coli* 0157:H7 in Water and Food

A technique has been developed for the direct detection of *E. coli* 0157:H7 in water
and food samples[5]. Immunomagnetic separation (IMS) is employed, using
magnetisable particles coated with an anti-O157 polyclonal antibody, followed by
incubation of the captured cells with cyanoditolyl tetrazolium chloride (CTC)
incubation to determine respiratory activity. Counter-staining with a specific
fluorescein-conjugated anti-O157 antibody (FAb) following CTC incubation was
used to allow confirmation and visualization of bacteria by epifluorescence
microscopy. *E. coli* 0157:H7 were used to inoculate fresh ground beef mince
(<17% fat), sterile 0.1% peptone, or water.
       The inoculated meat was diluted with an enzyme detergent buffer and
homogenized in a stomacher, followed by incubation with superparamagnetic beads

coated with anti-O157 specific antibody. After IMS, cells with magnetic beads attached were stained with CTC, then an anti-O157 FITC conjugate, and filtered for microscopic enumeration or solid- phase laser cytometry. The ChemScan RDI was used to detect the FITC fluorescence from the antibody, and CTC-formazan crystals in respiring cells were detected by microscopic validation following laser scanning. Enumeration by laser scanning permitted detection of ca.10 CFU per g of ground beef or < 10 per ml of liquid sample.

With inoculated peptone, regression results for log-transformed respiring FA positive counts of cells recovered on beads vs sorbitol-negative plate counts in the inoculum were: intercept=0.67; slope=0.88; r^2=0.98 (n=24). Comparable results for inoculated minced beef were: intercept=1.06; slope=0.89; r^2=0.95 (n=13). Within 5-7 h, the IMS/CTC/FAb method detected greater numbers of *E. coli* O157 cells than were detected by plating. The results show that the IMS/CTC/FAb technique with enumeration by either fluorescence microscopy or solid phase laser scanning cytometry gave results that compared favorably with plating following IMS.

IV Total Metabolically Active Bacteria

Burkholderia cepacia cultures which had been exposed to microgravity aboard NASA Shuttle Atlantis, ground and 1xg centrifuge controls were enumerated withR2A agar plate counts and by the use of the viability stain ChemChrome V3 (Chemunex) followed by ChemScan solid phase cytometry on return to earth[6]. ChemChrome V3 is a substrate which detects esterase activity by the cleavage of fluorescein diacetate and release of free fluoroscein within the cell. The samples were filtered through a black polycarbonate 0.2 μm pore-size membrane, incubated on a pad saturated with ChemChrome V3 for 30 min at 30°C, then scanned and validated microscopically. cultures were grown in sterile reagent grade water (Milli-Q, Millipore Corporation), 10% tryptic soy broth, or an iodine solution (ca. 2 mg/l iodine).

The cultures grew to 10^7-10^8 cfu/ml in 10% TSB, 10^5-10^6 cfu/ml in water, and 10^4-10^5 cfu/ml in iodine. Whether they had been subjected to microgravity or 1xg conditions, the ChemScan RDI results correlated well with the R2A agar plate counts performed in the laboratory. This experiment demonstrates the potential use of the ChemScan and ChemChrome substrate as a surrogate for the R2A agar heterotrophic plate count. The scanning method can be completed, including sample filtration and incubation, within 1 h.

VI Conclusions

Solid Phase Scanning laser cytometry provides a very convenient, sensitive and rapid method for the enumeration of fluorescently stained bacteria. It can be used for the direct detection of metabolically active bacteria in water samples. Using a fluorescein diacetate substrate (ChemChrome V3), results are equivalent to the heterotrophic plate count. Coliform bacteria and *E. coli* in water samples can be enumerated by pre-incubation and staining with fluorescent substrates which detect β-galactosidase in coliforms and β-glucuronidase in *E. coli*. In addition, Verocytotoxic *E. coli* O157:H7 in food and water can be concentrated by immunomagnetic separation followed by detection of metabolic activity and fluorescent antibody identification.

VII Acknowledgements

We thank Susan Broadaway and John Lisle, Montana State University, and staff of LigoCyte Pharmaceuticals Inc. (Bozeman, Montana, USA) for their excellent scientific advice and technical support. This research was supported financially by the U.S. National Aeronautics and Space Administration, U.S. National Institutes of Health, European Space Agency, and the Flemish Centre for Water Research. Chemunex (Maisons-Alfort, France) provided materials and advice.

VIII References

1. G.A. McFeters. 1990. Enumeration, occurrence, and significance of injured bacteria in drinking water. p. 478-492 *In* G.A. McFeters (ed.), Drinking water microbiology: progress and recent developments. Spring-Verlag New York, Inc., New York.
2. K. Mignon-Godefroy, J.-G. Guillet and C. Butor. 1997. Solid phase cytometry for detection of rare events. Cytometry 27:336-344.
3. H. Nelis. 1999. Enzymatic method for detecting coliform bacteria or *E. coli*. U.S. Patent 5,861,270. January 19, 1999.
4. S. Van Poucke and H. Nelis. 1999. A 3.5-h test for the determination of total coliforms and *Escherichia coli* in drinking water using two-stage enzymatic detection methodology and solid phase cytometry. Poster P9, International Conference on Rapid Detection Assays for Food and Water, 15-17 March 1999, Royal Society of Chemistry.
5. B.H. Pyle, S.C. Broadaway and G.A. McFeters. 1999. Sensitive detection of *Escherichia coli* O157:H7 in food and water using immunomagnetic separation and solid-phase laser cytometry. Appl. Environ. Microbiol. (In press.)
6. B.H. Pyle, G.A. McFeters, S.C. Broadaway, C.K. Johnsrud, R.T. Storfa and J. Borkowski. 1999. Bacterial growth on surfaces and in suspensions. *In* ESA SP-1222, Biorack on Spacehab. Biological Experiments on Shuttle to MIR Missions 03, 05, and 06. (In Press.)

Rapid Same-Day Microcolony Enumeration Technique For *E. coli* In Drinking Water

D.P. Sartory[1], A. Parton[2] and C. Rackstraw[2]

[1] Quality & Environmental Services, Severn Trent Water, Welshpool Road, Shelton, Shrewsbury, SY3 8BJ, U.K.
[2] Genera Technologies Ltd., Lynx Business Park, Fordham Road, Newmarket, Cambridgeshire, CB8 7NY, U.K.

1 INTRODUCTION

The enumeration of *Escherichia coli* and the related coliform bacteria has been the key bacteriological parameter monitored in the assurance of supply of safe drinking water. The basic techniques for the enumeration of *E. coli* were developed at the turn of the century (Sartory and Watkins, 1999). These were broth based and depended upon the calculation of a Most Probable Number (MPN) of the target organisms present. This remained the key method until the introduction of membrane filter techniques in the 1950s (Windle Taylor *et al.*, 1953). Currently, standard techniques based upon membrane filtration for the enumeration of *Escherichia coli* from drinking water take 18 to 24 hours to produce a presumptive result, with up to a further 24 hours for confirmation (Anon., 1994). The use of chromogenic and fluorogenic substrates for the detection of the β-glucuronidase enzyme has allowed greater confidence in enumerating confirmed counts of *E. coli* within 18 hours by both membrane filtration (Sartory and Howard, 1992) and MPN (Fricker *et al.*, 1997). More meaningful protection of public health, however, would be achieved if results of assays for *E. coli* were available as soon as possible after collection of samples, preferably within the working day to allow remedial action to be taken if necessary.

Since the mid 1970s there have been rapid developments in genetic, immunological and similar techniques, with the potential of allowing rapid same-day detection and, in some cases, enumeration of a wide range of bacteria. The use of polymerase chain reaction techniques (PCR) for detecting *E. coli* has been proposed by a number of researchers, but key problems exist relating to the detection of dead cells and inhibition by some sample constituents limiting its application for direct detection and enumeration of *E. coli* from water (Fricker and Fricker, 1997), although it may be useful for rapid confirmation of culturally grown isolates. Proposed immunological methods similarly have problems with sensitivity, cross reactions and detection of injured and dead cells (Sidorowicz and Whitmore, 1995). Additionally, these methods would be unlikely to correlate with current cultural methods, upon which regulatory and health responses are based.

The drawbacks associated with these potentially rapid methods makes the use of a cultural approach still attractive, provided that a reliable result can be obtained within a practical timeframe that can be regarded as sufficiently rapid to merit its adoption by water suppliers and regulators. With current cultural methods the use of less harsh media and enzyme specific substrates has allowed significant improvements in recoveries and identification of target bacteria but, although the time for confirmed results has been markedly reduced (e.g. from 4 days to 18 hours for *E. coli* and coliforms), there is still a significant time gap between collection of the sample and the availability of the result. Ideally for the water supply industry results should be available as soon as possible on the same day and early enough to allow immediate remedial actions.

2 CONCEPTS FOR SAME-DAY ENUMERATION

The advantages with the current methods for enumerating *E. coli*, especially those utilising specific substrates, are that they are relatively simple to perform and do not require sophisticated laboratory equipment. Unfortunately they are also labour intensive and require at least overnight incubation. With appropriate manpower resourcing, however, a relatively large number of samples (typically 200 - 500) can be analysed daily with limited laboratory facilities and junior grade analysts. For a method to be able to provide results on the day of sample collection it would need to meet some key criteria with respect to current capabilities and legislation before adoption. These are (Sidorowicz and Whitmore 1995):-

      i)   Results to be available in 6 - 8 hours or less;
      ii)   Results to be quantitative; and
     iii)   Close comparability with existing methods.

     Standards for *E. coli* in drinking water are legally defined and there are substantial databases relating to the performance to these standards and, with a record of virtually no bacteriological waterborne outbreaks attributable to treated water, any change in methods that produces a significant increase in positive samples will generate concerns regarding the interpretation and significance of such results. Water companies may see their water quality performance seemingly deteriorate in respect to previous results due to the improved methodology but, arguably, without any increase in health risk.

     There are some key attributes that a rapid same-day method for *E. coli* would need before becoming acceptable to microbiologists. These primarily relate to method performance, costs and support from the manufacturer (Sartory and Watkins, 1999):-

i)   Sensitivity and specificity - the system must be able to accurately detect the target organisms from high concentration of background and competing organisms. Sensitivity must be at least 1 organism per 100 ml. False-positives and false-negatives must be as close to zero as possible.

ii)   Speed - For operational actions to be undertaken within social hours (say before 8.00 pm) then results need to be available by late afternoon and preferably earlier. Since there may be restrictions on how early in the morning samples can be taken, then the time between collection of the sample and availability of results needs to be minimised, leaving a maximum of about eight hours for growth and detection.

iii)   Throughput - For the bigger water companies microbiology laboratories typically have a daily sample throughput of 200 - 500 samples. Systems offering same-day results need to be able to handle this throughput without the need for banks of expensive detection equipment.

iv)   Non-destructive - There will always be a requirement to be able to subculture isolated bacteria for further study. The system must allow easy retrieving of such isolates.

v)   Analytical skills - Any system developed should not be so sophisticated as to require analysts with degrees or doctorates. Automated or semi-automated equipment may offer advantages in ease of use by analysts with basic or no microbiological training. Problem solving, however, may require more detailed knowledge.

vi)   Costs - The cost of initial purchase of the equipment and subsequent costs for reagents and materials must to be commensurate with an acceptable cost per test in relation to current analytical costs. Expensive equipment may need to achieve very high throughputs per instrument compared to alternative cheaper equipment with lower throughputs per instrument.

vii)   Manufacturer's reliability and technical service - With a legal responsibility on water companies to achieve stipulated sample and analysis frequencies the manufacturers of equipment supplied to the water industry must have a very high level of reliability and technical back-up. The more complicated the equipment the more the laboratory is reliant upon the manufacturer's technical services.

3 THE MICROCOLONY APPROACH

Taking the above criteria and constraints into account, the detection of microcolonies after a short period of growth offered the greatest likelihood of being acceptable to bacteriologists and regulators. This can be achieved with a short period of culture, possibly boosted by the addition of growth supplements, to produce a microcolonies of the target organisms. This at least demonstrates viability and, if growth is on selective media, limits background interferences. Growth of target organisms into microcolonies also suggests that the samples would also have been positive had conventional culture continued. Providing the target signal is large enough detection could be carried out simply using laser scanning or high sensitivity cameras, either with pre-staining the colony with a suitable monoclonal antibody or via a viability or substrate based marker. Scanning of membranes would need to be relatively fast, as it is with conventional microbiology. A scanning speed of 60 seconds per sample for 500 samples would, for a single instrument, require 8.3 h to complete, abrogating much of any advantages over current conventional culture.

A variety of methods can be employed to detect microcolonies of *E. coli* or other organisms, some of which have been reviewed by Sartory and Watkins (1999). DNA binding stains, such as acridine orange and DAPI, or viability dyes, such as 5-cyano-2,3-ditolyl tetrazolium chloride (CTC), can be used for non-specific signalling of microcolonies. Rodrigues and Kroll (1988) described the detection of microcolonies of a range of micro-organisms from foods, including *E. coli* and coliforms, grown on selective media and realised with acridine orange. Microcolonies of laboratory grown *E. coli* could be detected after 3 h incubation at 30°C on enriched lauryl sulphate aniline blue agar. Newby (1991) used the same approach to enumerate microcolonies of *Pseudomonas cepacia* inoculated into sterile pharmaceutical grade water, with incubation on TSA at 31°C for 4 h, achieving a sensitivity of 10 cfu/ml and results within the working day. Viability staining could be combined with an immunofluorescence stain to allow specific detection. A combined immunofluorescence/CTC method for the detection of single cells of *E. coli* O157:H7 in water has been described, and similar procedures applied to *Salmonella typhimurium* and *Klebsiella pneumoniae* (Pyle *et al*. 1995). This approach using monoclonal antibodies against *E. coli* together with a suitable viability dye could provide a rapid means of detecting microcolonies. Such a system, however, would require monoclonals with a high specificity to ensure detection of the target organisms only. This may be achievable with specific strains of *E. coli*, but may be difficult when trying to encompass all possible, or just the most commonly isolated, strains. An alternative approach was adopted by Meier *et al*. (1997) for the detection of microcolonies of species of *Enterococcus* (of which the microcolonies of some strains could be visualised after four to five hours incubation on a non-selective medium) by employing fluorescently labelled rRNA targeted DNA probes. Specificity of detection could also be achieved through the use of genetically modified *lux* bearing bacteriophage constructs, which have been derived for a number of specific hosts including *E. coli* (Kodikara *et al*. 1991). All of these approaches, however, would require significant multi-step process handling for reaction prior to detection of the microcolonies.

The use of enzyme specific substrates in specific or non-specific media offers a simple and potentially clean system for the detection of microcolonies. Well established chromogens and fluorogens are available for the detection of β-glucuronidase and have been employed in a wide range of media for the detection of *E. coli* (Sartory and Watkins, 1999).

4 DEVELOPMENT OF A MICROCOLONY DETECTION METHOD FOR *E. COLI* IN DRINKING WATER

We have developed a system, based upon the detection of microcolonies, by which enumeration of *E. coli* from a large number (200 - 600) drinking water samples can be achieved within six hours of collection of the samples. To achieve this capability a number of analytical requirements, logistical issues and practicalities had to be addressed, and technical problems resolved. The system developed consists of processing samples at the time of collection, incubation during transport to the laboratory and specific detection and enumeration of target microcolonies at the laboratory.

4.1 On-site Sample Filtration

The key issue for filtering water samples at the point of collection is to ensure filtration of the desired volume (100 ml) within acceptable tolerance limits and in such a fashion as to ensure the sample is not contaminated during processing. There are a number of filtration systems commercially available that are typically used for environmental monitoring. These were, however, considered not protected enough for routine filtering of regulatory drinking water samples. Consequently we have devised an enclosed filtration unit (40 mm diameter x 28 mm deep) containing a 25 mm diameter 0.45 μm pore size black cellulose acetate filter (Sartorius) supported on an absorbent pad and filter disc, an ampoule containing growth medium and fitted with removable connection tubes. The sample is collected in a specific sample bottle to a marked level, after which the original cap is replaced with a sterile cap fitted with an internal draw-off tube and closable port.

The cap allows withdrawal of 100 ml ± 5 ml of sample. There are no recommended tolerances for sample volume filtered in UK guidance (Anon., 1994), but the USEPA has suggested that analysts should achieve a ± 2.5% accuracy in volumes for membrane filtration (Geldreich, 1975). This proved a tight tolerance to achieve. The filtration unit is connected to the cap port and a vacuum pump and reservoir in a carry case and the desired sample volume filtered, after which the medium ampoule is broken via a screw mechanism, releasing the medium into the absorbent pad. The whole operation takes less than five minutes and at no time is the sample, filter or medium exposed to potential contamination. Field trials have shown that the system can be easily used in the open back of a van under adverse conditions without contamination occurring.

4.2 Incubation During Transport

It is essential that if incubation is to be undertaken during transportation in a van to the laboratory that the incubation temperature is maintained within accepted tolerance limits (± 0.5°C). The incubator must also be physically and electronically robust. We tested two commercially available incubators, both being small chambers, one fan assisted. Neither were able to maintain the required temperature, as typically the whole chamber volume was replaced with cold air each time a sample was put in for incubation, and both were unable to restore the desired incubation temperature during a typical series of sampling events. Consequently an incubator had to be designed. Aluminium heating blocks with drilled holes for tubes, with good temperature control, have been commercially available for many years. Adopting this approach in a suitably designed housing has proved successful for the small filtration units, with a temperature variation of ± 0.2°C being achieved. These field incubators can be powered from the mains via an adapter, from a car/caravan battery or from an internal battery, making it very flexible for use under a variety of scenarios. The incubators are capable of incubating up to 27 samples (sufficient for a typical sampling run or incident samples) and are fitted with temperature data loggers which can be downloaded to a computer at the end of each day for quality assurance records. As the incubators need to be carried a weight limit of less than 10 kg was imposed.

4.3 Detection and Enumeration of Microcolonies

There are a number of approaches for generating a detection signal from microcolonies of *E. coli*. In order to keep processing of the sample at the laboratory simple and rapid a combined fluorogenic viability stain/specific enzyme substrate (for β-glucuronidase) approach was adopted. Microcolonies of *E. coli* are grown on a modification of membrane-lauryl sulphate broth (the UK standard medium, Anon. 1994) containing 4-methylumbelliferyl-β-D-glucuronide (MUG), a well established fluorogen for the detection of *E. coli*. MUG is readily metabolised by *E. coli* and the cleaved product generates a good fluorescent signal. As the incubation period for microcolony growth is relatively short, diffusion of the fluorescent product is limited and does not cause the problems encountered when MUG has been used for conventional macrocolony filtration methods (Sartory and Howard, 1992). An alternative red fluorescing fluorogen, 2-dodecylresorufin-β-D-glucuronide (Molecular Probes Inc.), which also produces a good fluorescent signal, was tested but uptake was poor and could not be significantly improved by increasing membrane permeabilisation with polymixin B. Microcolonies of *E. coli* can be detected after 5.5 h incubation over a range of temperatures from 35°C to 42°C. Whilst greater expression of β-glucuronidase was achieved at 42°C, better recovery of stressed *E. coli* was achieved at 35°C to 37°C. In our approach the initial identification of microcolonies is through a fluorescent signal generated by metabolisation of CTC. After incubation for 5.5 h the membrane from the filtration unit is removed and placed upon a pad soaked in CTC solution and incubated at 42°C for 0.5 h. This stains the microcolonies with the viability marker and also increases the metabolism of MUG increasing the β-glucuronidase signal. The microcolonies can then be enumerated under standard epifluorescence microscopy for CTC with *E. coli* being confirmed by switching to the detection of the fluorescence caused by the cleaving of MUG. We are currently developing a computer controlled epifluorescence system for the semi-automated enumeration of microcolonies, which is based upon the use of image analysis software that discriminates the microcolonies on the bais of size, shape and intensity of fluorescence. As the approach adopted is non-destructive the microcolonies can be picked off, if required, for confirmation tests. Recoveries of microcolonies of *E. coli* grown under these conditions equivalent to recoveries by conventional membrane filtration using m-LGA (Sartory and Howard, 1992) have been achieved.

5 ACKNOWLEDGEMENTS

This paper is published with the permission of Severn Trent Water Ltd and the opinions expressed are those of the authors and do not necessarily reflect those of the company. We are indebted to the many people who have been involved in various parts of this work, particularly Malcolm Bird, David Dawson, Kaye Power, Adele Pritchard, Mark Field and Stuart Curbishley.

6 REFERENCES

Anon. (1994) *The Microbiology of Water 1994: Part 1 - Drinking Water*. Reports on Public Health and Medical Subjects No. 71. Methods for the Examination of Water and Associated Materials. London: HMSO.

Fricker E.J. and Fricker C.R. (1997) Use of the polymerase chain reaction to detect *Escherichia coli* in water and food. In *Coliforms and E. coli: Problem or Solution?* (ed. Kay D. and Fricker C.), Cambridge:The Royal Society of Chemistry. pp 12-17.

Fricker E.G., Illingworth K.S. and Fricker C.R. (1997) Use of two formulations of Colilert and QuantiTray™ for assessment of the bacteriological quality of water. *Water Research* **31**, 2495-2499.

Geldreich E.E. (1975) *Handbook for Evaluating Water Bacteriology Laboratories*, 2nd Edition. U.S Environmental Protection Agency publication EPA-670/9-75-006. Cincinnati:USEPA.

Kodikara C.P., Crew H.H. and Stewart G.S.A.B. (1991) Near on-line detection of enteric bacteria using *lux* recombinant bacteriophages. *FEMS Microbiology Letters* **83**, 261-266.

Meier H., Koob C., Ludwig W., Amann R., Frahm E., Hoffmann S., Obst U. and Schleifer K.H. (1997) Detection of enterococci with rRNA targeted DNA probes and their use for hygienic drinking water control. *Water Science and Technology* **35**, 437-444.

Newby P.J. (1991) Analysis of high-quality pharmaceutical grade water by direct epifluorescent filter technique microcolony method. *Letters in Applied Microbiology* **13**, 291-293.

Pyle B.H., Broadway S.C. and McFeters G.A. (1995) A rapid, direct method for enumerating respiring enterohemorrhagic *Escherichia coli* O157:H7 in water. *Applied and Environmental Microbiology* **61**, 2614-2619.

Rodrigues U.M. and Kroll R.G. (1988) Rapid selective enumeration of bacteria in foods using a microcolony epifluorescence microscopy technique. *Journal of Applied Microbiology* **64**, 65-78.

Sartory D.P. and Howard L. (1992) A medium detecting β-glucuronidase for the simultaneous filtration enumeration of *Escherichia coli* and coliforms from drinking water. *Letter in Applied Microbiology* **15**, 273-276.

Sartory D.P. and Watkins J. (1999) Conventional culture for water quality assessment: is there a future? *Journal of Applied Microbiology, Symposium Series* (in press).

Sidorowicz, S.V. and Whitmore, T.N. (1995) Prospects for new techniques for rapid bacteriological monitoring of drinking water. *Journal of the Institute of Water and Environment Management* **9**, 92-98.

Windle Taylor E., Burman N.P. and Oliver C.W. (1953) Use of the membrane filter in the bacteriological examination of water. *Journal of Applied Chemistry* **3**, 233-240.

IMMUNOMAGNETISABLE SEPARATION FOR THE RECOVERY OF *CRYPTOSPORIDIUM* sp. OOCYSTS

[1]C.A. Paton, [2]D.E. Kelsey, [2]E.A. Reeve, [3]K. Punter, [2]J.H. Crabb and [1]H.V. Smith.

[1]Scottish Parasite Diagnostic Laboratory, Stobhill NHS Trust, Springburn, Glasgow G21 3UW, [2]ImmuCell Corporation, Portland, Maine 04103, USA, and [3]West of Scotland Water Authority, Customer & Environmental Services, Leven House, Balmore Road, Glasgow G22 6NU.

1 INTRODUCTION

The protozoan parasite *Cryptosporidium parvum* can cause diarrhoeal disease in human beings and other mammals, infection resulting from the ingestion of environmentally robust oocysts excreted in the faeces of infected hosts. Du Pont *et al.*, (1995)[1] demonstrated that 30 oocysts of a bovine *C. parvum* isolate caused infection in 20% of healthy human volunteers (ID_{50} = 132 oocysts). Waterborne cryptosporidiosis can infect large numbers of consumers from a single contamination event and in the last 13 years at least 19 waterborne outbreaks, affecting an estimated 427,100 individuals[2] have been documented. The use of contaminated potable water in food production and the globalisation of food sources also enhance the likelihood of foodborne transmission.[3] Three foodborne outbreaks of cryptosporidiosis have been documented[3] and *Cryptosporidium* oocysts have been detected as surface contaminants on a variety of vegetables. Rapid, sensitive methods are required to determine the occurrence of oocysts in both water concentrates and in/on food matrices.

Methods for the isolation and detection of *Cryptosporidium* in water and associated matrices are time-consuming, tedious and inefficient. Flotation methods for oocyst clarification are inefficient and do not separate oocysts effectively from contaminating particulates. Immunomagnetisable separation (IMS) has proved useful in the selective enrichment of *Cryptosporidium* oocysts from seeded and environmental samples[4,5,6,7] and has potential for recovering oocysts from turbid food suspensions. IMS kits use anti-*Cryptosporidium* oocyst wall-reactive monoclonal antibodies (mAbs) which can vary in their isotype and affinities. Variable recovery efficiencies have been reported,[6,7] influenced by factors such as seeding dose and sample turbidity. This study compared the ability of two commercially available IMS kits to recover *C. parvum* oocysts from clean and turbid water concentrates.

2 MATERIALS AND METHODS

2.1 Sources of *C. parvum* oocysts: Six sources of purified *Cryptosporidium* oocysts were used. *1*: Oocysts of the Iowa (bovine) isolate purchased from University of Arizona, USA. *2*: A pool of oocysts from source 1. *3*: UCP strain passaged in calves at Tufts University

School of Veterinary Medicine, USA. *4*: GCH-1 strain passaged in calves (provided by Dr. G. Widmer, Tufts University School of Veterinary Medicine, USA). *5*: GCH-2 strain passaged in pigs (provided by G. Widmer). *6*: Iowa (bovine) isolate purchased from Pleasant Hill Farm, Idaho, USA.

2.2 Sources of water concentrates: Turbid concentrates were prepared by washing yarn wound filters (ICR type or Cuno DPPPY) through which water from Portland, Maine, Omaha, Nebraska, USA or west of Scotland, UK had been filtered.

2.3 IMS kits: Two commercially available IMS kits, Crytpo-Scan IMS; ImmuCell Corporation, Portland, Maine, USA (kit A) and Dynabeads anti-*Cryptosporidium*; Dynal A.S., Oslo, Norway (kit B), were assessed for their ability to recover *C. parvum* oocysts seeded into reverse osmosis (RO) water and turbid water concentrates (57-15,000 nephelometric turbidity units (NTU)). The kits were used according to manufacturers' instructions. In addition to the kit A format, a further format of the ImmuCell IMS was assessed, whereby capture of bead-oocyst complexes by panning in a horizontal plane was replaced by capture on a vertical magnetic clip stand. The protocol was similar, except that bead-oocyst complexes were magnetised for a period of 3-5 min (see Figures 4&5).

2.4 Enumeration of *C. parvum* oocysts: Purified stock suspensions of *C. parvum* oocysts were enumerated by haemocytometer and diluted in RO water. Aliquots (10 x 50 μL) of oocyst suspensions were applied onto four-well microscope slides, air dried, methanol fixed (5 min) and overlaid with 50 μL of fluorescein isothiocyanate-conjugated anti-*Cryptosporidium* monoclonal antibody (FITC-mAb; Waterborne Inc. New Orleans, USA; ImmuCell Corporation). Samples were incubated in a humidified chamber at 37°C for 30 min, washed twice in phosphate buffered saline (PBS; 150mM, pH 7.2) and once in PBS with 4'6-diamidino-2-phenylindole (DAPI; 50 mL PBS : 10 μL 2 mg mL^{-1} DAPI). One drop (50 μL) of RO water was added to each well and allowed to air dry in the dark. Samples were mounted in glycerol:PBS (60:40 v/v) containing 2% of the antifadant DABCO, a coverslip applied and examined using epifluorescence microscopy.

2.5 Microscopy: Oocysts were enumerated using an Olympus BH-2 or a Nikon Diaphot TMD epifluorescence microscope equipped with a blue filter block for visualisation of FITC (480nm excitation; 520 nm emission), an ultra violet filter block for visualisation of DAPI (375nm excitation; >420nm emission) and Nomarski DIC optics for visualisation of internal contents.

2.6 Seeding of samples: Prior to seeding, 8-12 replicates of diluted oocyst suspensions were applied onto welled microscope slides, air dried, methanol fixed, labelled with FITC-mAb and enumerated as above. Between 10 and 200 oocysts were seeded into 10 mL volumes of RO water and turbid water concentrates (see figures).

2.7 Field Study: Fifteen final water samples from a distribution outlet suspected of containing *Cryptosporidium* oocysts were filtered (range 222-1573L) and samples were processed according to either the UK provisional recommended method[8] or manufacturers' instructions. Fluorescence activated cell sorting (FACS)[8] and IMS (kit A) were used to clarify samples.

3 RESULTS

3.1 Effect of increased separation time and bead concentration (kit A): Using 50 µL of Crypto magnetic beads and rotating the sample on the magnetic panning device (MPD) for 2 min, produced a mean % recovery of 43.5 ± 23.4 (n = 6) from samples seeded with 13.8 ± 4.0 oocysts (n = 12). Increasing the volume of beads to 100 µL and allowing the sample to rotate for 5 min on the MPD, produced a mean % recovery of 56.1 ± 20.5 (n = 6) from samples seeded with 10.7 ± 1.5 oocysts (n = 12).

3.2 Assessment of kit performance using various *C. parvum* isolates: The number of oocysts recovered from clean and turbid samples using kits A and B varied from isolate to isolate. Overall, higher recoveries were achieved from clean and turbid water samples (≤0.5 mL packed pellet) using kit A than with kit B in 13 out of 14 occasions. Kit A performed well with all *Cryptosporidium* isolates with one exception. Recoveries from clean water samples spiked with 1 batch of the Iowa isolate (source 2) were poor (mean 11.8%). In this instance, although oocysts were captured, their release, following acid desorption, was ineffective. In contrast, kit B performed well with the Iowa isolate (mean recovery 74.1%), but less well with all other isolates (Figures 1 and 2).

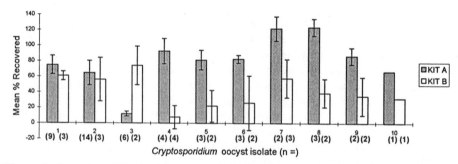

Figure 1: *Recovery of C. parvum oocysts from clean water samples seeded with 100-200 oocysts.*

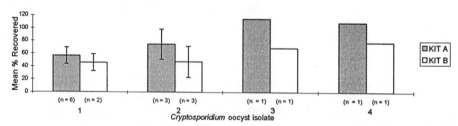

Figure 2: *Recovery of C. parvum oocysts from 57-5400 NTU water concentrates seeded with 100-200 oocysts.*

3.3 Recovery of *C. parvum* oocysts from high turbidity water concentrates - a comparison of two commercially available kits: Water concentrates prepared from a west of Scotland source (15,000 NTU; 2 mL packed pellet) were seeded with between 44.5 ± 5.2 oocysts (n = 8) and 48.1 ± 9.7 (n = 16) oocysts. A mean % recovery of 29.3 ±

13.9% (n = 10) was achieved from samples using kit A compared to a mean % recovery of 8.3 ± 4.4% (n = 10) from samples using kit B (Figure 3).

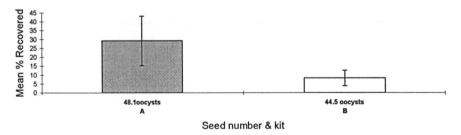

Figure 3: *Recovery of C. parvum oocysts from 15,000 NTU water concentrates seeded with 44.5-48.1 oocysts.*

3.4 Recovery of *C. parvum* oocysts using kit A from turbid (2900-5000 NTU) water concentrates - comparison of reaction vessels and magnetic bead size: Water concentrates prepared from water filtered around Portland, Maine (2900 NTU) and the Omaha region of Nebraska (5000 NTU) were seeded with between 89 and 133 oocysts. Three reaction vessels (MPD, polypropylene tubes and Leighton tubes) and 2 lots of magnetic beads (1 μm & 4 μm dia.) were used to recover oocysts from these water concentrates (0.5 mL packed pellet). In both water types, higher recovery efficiencies (60.5-68.4%) were achieved using 4 μm dia. magnetic beads and Leighton tubes (Figures 4 & 5).

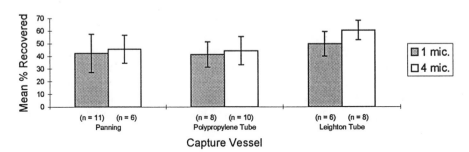

Figure 4: *Recovery of C. parvum oocysts (seed = 89-133 oocysts) from 2900 NTU water concentrates: comparison of reaction vessels and magnetic bead size.*

3.5 Field Study: Twelve (80%) of the 15 samples tested were positive for oocysts. Seven (58.3%) of the 12 samples were positive by FACS (range: 0.003-0.045/L) and 10 (83.3%) were positive by IMS (range: 0.005 - 0.045/L).

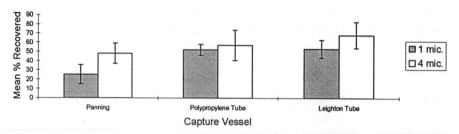

Figure 5: *Recovery of C. parvum oocysts (seed = 126-133 oocysts) from 5000 NTU water concentrates: comparison of reaction vessels and magnetic bead size (n = 3).*

4 DISCUSSION

Water concentrates from both UK and US sources, were seeded with low (10-200) densities of *C. parvum* oocysts and the recovery efficiencies of two commercially available IMS kits assessed. Altering the parameters of panning time and bead volume for kit A resulted in >10% increase in oocysts recovered. Doubling the volume of beads and rotating the sample for 5 min on the MPD increased mean % recovery from 43.5 ± 23.4 (n = 6) to 56.1 ± 20.5 (n = 6). This modification was adopted for the remaining trials. Kit A performed well with all *Cryptosporidium* isolates except one batch of the Iowa isolate, prepared from 2 separate lots of oocysts. In contrast, kit B performed well with all Iowa isolates but less well with all other isolates (Figure 1). Consistently higher recoveries were achieved from seeded water concentrates using kit A (Figures 1-3). Furthermore, kit A recovered more oocysts from high turbidity samples than kit B. In a direct comparison of both kits to recover oocysts from 15,000 NTU concentrates, higher recovery efficiencies were obtained using kit A. The kit B manufacturer's protocol states that the packed pellet volume of samples should not exceed 0.5 mL and that samples with a pellet volume >0.5 mL should be sub-divided so that each sub-sample contains ≤0.5 mL packed pellet. Packed pellets of 0.5 mL can be generated when 10L of raw water are filtered. The packed pellets of the 15,000 NTU samples were 2 mL and samples were not subdivided into 0.5 mL aliquots. IMS was performed on the 2 mL packed pellet samples and may explain the exceptionally low recoveries achieved using kit B. Subdivision of the samples would have resulted in a time consuming and costly approach since four IMS procedures would be required for each 15,000 NTU sample. Packed pellets >0.5 mL can result when 1000L of treated water are filtered and may well be encountered when the new UK *Cryptosporidium* regulation[9] is enacted. When the proposed regulation was promulgated, IMS was identified as the oocyst concentration step.

Both oocyst isolate and water concentrate composition influenced kit performance. A variety of factors, including antibody affinity and isotype, oocyst epitope expression and the physico-chemical environment can affect epitope-paratope interaction. Interaction is most stable with high affinity antibodies. Most commercially available *Cryptosporidium* mAbs are IgM isotype (including kit B), whereas the kit A mAb is an isotype switched, affinity matured IgG_3. Affinity matured paratopes normally bind epitopes tighter than non-isotype switched paratopes. In IMS, antibody capture and release are trade-offs, in that paratopes which bind strongly to their epitopes are less easy to dissociate. Lower affinity paratopes can bind oocyst epitopes effectively in favourable conditions, such as lower turbidity concentrates, but less effectively in adverse conditions, such as the higher turbidity concentrates tested. Higher affinity mAbs bind oocyst epitopes in more adverse conditions, such as those described (Figures 1-3). Therefore, local water compositions can have a major influence on IMS performance with factors such as pH, turbidity and divalent cations affecting antibody binding.

Previous comparisons of commercially available ImmuCell and Dynal IMS kits obtained higher recoveries with the Dynal kit[6,7] however, the ImmuCell kits used for much of the comparisons used non-releasable beads which hamper identification because the beads mask oocysts. In addition, few replicates were performed, which reduces statistical validity. On the basis of our data, kit A was chosen for recovering oocysts from a source known to have sporadic *Cryptosporidium* contamination. Comparison with FACS analysis indicates that IMS is more effective, recovering oocysts from 83.3% of samples compared with 58.3%.

Commercial IMS kits can provide an effective method for concentrating waterborne *Cryptosporidium* oocysts however, the effects of antibody isotype, oocyst isolate and water concentrate composition which can influence oocyst recoveries and kit performance, must be determined empirically for each kit. Similar factors will be significant in food matrices. IMS is a useful technique and can be readily allied to *in vitro* culture and immunological/molecular methods (PCR-RFLP) for the specific enumeration and species identification of small numbers of *Cryptosporidium* oocysts.[3]

REFERENCES

1. H. Du Pont, C. L. Chappell, C. R. Sterling, P. C. Okhuysen, J. B. Rose, and W. Jakubowski, *N. Engl. J. Med.*, 1995 **332**, 855.
2. H.V. Smith and J. B. Rose, *Parasitol. Today*, 1998, **14**, 14.
3. R.W.A. Girdwood and H.V. Smith (1999). *Cryptosporidium*. "In: Encyclopaedia of Food Microbiology" (eds. R. Robinson, C. Batt and P. Patel) Academic Press, London & New York. In press.
4. A.T. Campbell and H.V. Smith, *Wat. Sci. Tech.*, 1997, **35**, 397.
5. A.T. Campbell, B. Grøn and S.E. Johnsen (1997). In: "Proceedings of the 1997 AWWA International Symposium on Waterborne *Cryptosporidium*". (Eds. C.R. Fricker, J.L. Clancy, and P.A. Rochelle). American Water Works Association, Denver, Co, USA. pp. 91-96.
6. P.A. Rochelle, R. De Leon, A. Johnson, M. H. Stewart, and R. L. Wolfe, *Appl. Environ. Microbiol.*, 1999, **65**, 841.
7. Z. Bukhari, R.M. McCuin, C.R. Fricker, and J.L. Clancy, *Appl. Environ. Microbiol.*, 1998, **64**, 4495.
8. Anonymous (1999). Methods for the examination of waters and associated materials. *Isolation and identification of Cryptosporidium oocysts and Giardia cysts in water* 1999. Environment Agency, HMSO, London. 44 pp.
9. Anonymous (1999). The water supply (water quality) (amendment) regulations 1999. HMSO.

USE OF A COMBINED IMS/LASER SCANNING PROCEDURE FOR THE SEPARATION AND SUBSEQUENT DETECTION OF *CRYPTOSPORIDIUM* OOCYSTS FROM WATER CONCENTRATES.

D.T.Reynolds, R.B.Slade and C.R.Fricker

Development Microbiology, Thames Water Utilities, Manor Farm Road, Reading, Berkshire, UK. RG2 OJN

1 INTRODUCTION

Methods for the isolation of *Cryptosporidium* oocysts have received much attention over recent years largely due to an increase in the number of reported waterborne outbreaks of cryptosporidiosis. Whilst a variety of techniques are being developed to detect oocysts from water concentrates the majority of laboratories rely upon the use of immunofluorescence assays. In such procedures samples are incubated with a fluorescein isothiocyanate conjugated *Cryptosporidium sp.* specific monoclonal antibody (FITC-mAb) prior to examination using epifluorescence microscopy. Although such assays may be used to directly examine water concentrates, the presence of particulate material which accumulates during the concentration procedure often interferes with this analysis. Conventional techniques employed to reduce levels of this debris involve flotation procedures or flow cytometry. Of these two techniques flow cytometry has proved most useful, however, requires highly trained personnel and is relatively slow. Recently, however, immunomagnetic separation (IMS) procedures have become available which have been reported to provide a reliable alternative to conventional separation techniques[1,2]. Using this technology magnetic beads coated with monoclonal antibody specific for *Cryptosporidium sp.* are used to isolate oocysts from water concentrates.

Following IMS, oocyst detection relies upon the examination of sample preparations using epifluorescence microscopy. This procedure is extremely time consuming and can cause operator fatigue. Consequently, a reliable automated procedure would be of considerable benefit. Laser scanning, using the Chem*Scan* RDI instrument, is a new technique which may be used for the rapid detection of fluorescently labelled *Cryptosporidium* oocysts. Using this method, organisms are captured by membrane filtration, labelled and the filter scanned with a laser inside the instrument. Subsequently, the signals generated undergo a sequence of computer analyses in order to identify events that are labelled organisms (scanning is complete within 3-4 minutes). A visual validation of all the Chem*Scan* results can then be made by transferring the membrane to an epifluorescence microscope that is fitted with a motorised stage. This stage, which is controlled by the Chem*Scan* RDI, can be driven to the location of each fluorescent event for visual confirmation of all results.

In this study we describe the application of the Chem*Scan* system for the rapid detection of *Cryptosporidium* oocysts after their isolation from water concentrates using IMS.

2 MATERIALS AND METHODS

2.1 Organisms

Cryptosporidium parvum oocysts (Harley Moon strain, NADC, Ames IA, US) were obtained from the Sterling Parisitology Department, University of Arizona (US).

2.2 Concentration procedure

Samples of river water (30L) were taken from various sites within the Thames Water Utilities catchment area. Water from each was processed in 10L volumes, using calcium carbonate flocculation/centrifugation.

2.3 Oocyst seeding

Aliquots (500µl packed pellet volume, 1050g, 10 minutes) of each water concentrate were added to Leighton tubes containing a 'seed dose' of 100 oocysts produced using flow cytometry. Three replicates of each water concentrate were prepared. In addition an unseeded control was also prepared.

2.4 Immunomagnetic separation

This procedure was based upon the Dynal IMS procedure (Dynal AS, Oslo, Norway). The volume in each Leighton tube was made up to 10 ml with de-ionised water. Following this dilution, 1ml of 10x SL Buffer A (Dynal), 1 ml of 10x SL Buffer B (Dynal) and 100µl of Dynabeads (Dynal) were added. Each sample was incubated on a rotating mixer at room temperature for 1 hour. Immediately after incubation, the paramagnetic beads were captured using a magnetic concentrator (MPC-1, Dynal) and the supernatant carefully poured to waste. The beads were re-suspended in 1ml of 1x SL buffer A and transferred to a 1.5ml Eppendorf tube where they were recaptured using a second magnetic particle concentrator (MPC-M, Dynal). After the resultant supernatant was discarded the beads were re-suspended in 100µl of 0.1M hydrochloric acid and incubated at room temperature for 5 minutes. After incubation, the beads were recaptured and the acid supernatant transferred to a clean Eppendorf tube. This acid incubation was repeated.

2.5 Oocyst permeabilisation

This procedure was performed to subsequently permit effective oocyst labelling with 4,6-diamidino-2-phenylindole (DAPI). The acidified supernatant (200µl) obtained from the IMS procedure was neutralised by the addition of 200µl of MOPS buffer (1M, pH7, Sigma). Each sample was then diluted with 400µl of ethanol (99.9%) and incubated in a water bath (80°C, 10 minutes). Oocysts within this final supernatant were then captured by membrane filtration (2µm pore size, Cycloblack polyester membranes) prior to incubation with FITC-mAb/DAPI.

2.6 Enumeration of *Cryptosporidium* oocysts

This was performed following instructions given by the 'ChemScan Detection of *Cryptosporidium* oocysts on a Membrane filter Kit' (Chemunex, Maison Alfort, France). Briefly, after oocyst capture, each membrane was carefully transferred onto a sintered glass vacuum support and rinsed with 500μl of a 1:5 dilution of ChemSol B12 (Chemunex). Subsequently, each membrane was carefully loaded on a droplet (100μl) of labelling solution and incubated in a humid chamber at 37°C for 1 hour. The labelling solution was prepared by mixing ChemID (Chemunex) which contains a *Cryptosporidium* specific FITC-mAb and CSH (Chemunex) which contains an antibody blocking reagent and DAPI as instructed. After incubation, each membrane was analysed with the Chem*Scan RDI* instrument following the instructions given by the manufacturer. After analysis, all fluorescent events on the membrane, which were identified by the Chem*Scan RDI* as potentially being an oocyst, were microscopically validated. This allowed a visual confirmation whether the fluorescent event detected was a labelled oocyst by assessing FITC-mAb/DAPI staining characteristics.

3 RESULTS

The recovery of oocysts from each matrix using IMS was found to be high (Figure 1) with a mean percentage recovery of 81.9% ± 3.32. Subsequent analysis of this recovery data demonstrated no significant difference between recoveries obtained from these different matrices (t-test, $p > 0.05$).

During the analysis of these IMS supernatants using the Chem*Scan* system a number of fluorescent events were detected on each membrane which were not confirmed as being labelled oocysts during the microscopic validation procedure (mean = 33.32 ± 18.38). These appeared to be either particulate debris or cellular material to which the FITC-mAb had non-specifically bound. Although these events interfered with the analysis, the microscopic validation system allowed rapid validation of the results (approximately 100 events in 3-4 minutes).

Figure 1. *Percentage recovery of oocysts using immunomagnetic separation.*

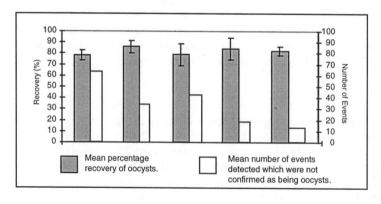

4 DISCUSSION

It is widely recognised that the methods currently used for the isolation of oocysts from water concentrates are highly inefficient[3,4]. In order to address the limitations of these methods we investigated the use of IMS. Results demonstrated a mean oocyst recovery of 81.9% with no significant difference between the recoveries obtained from the different matrices (p>0.05). Such high recoveries using this IMS kit have been previously reported for both fresh and aged oocysts[1,2]. In this series of investigations the number of FITC-mAb labelled oocysts present upon membrane filters was carried out using a novel laser scanning device (Chem*Scan*) which proved to be a highly reliable instrument for the enumeration of these organisms. Conventional techniques for the examination of IMS supernatants for the presence of oocysts rely upon the fixing of the sample upon microscope well slides prior to labelling with FITC-mAb and examination using epifluorescence microscopy. In our experience, the microscopic examination of each sample requires approximately 15-20 minutes with the time required increasing as subsequent preparations are analysed (this is due to operator fatigue). In contrast, examination of sample preparations using the Chem*Scan* system may be completed in approximately 5 minutes. In addition, during the manual microscopic examination of samples many laboratories routinely use a magnification of x200 as this allows samples to be analysed at a much faster rate than at higher magnifications. However, it is our experience that oocysts in environmental samples often do not stain efficiently with FITC-mAb and consequently can be difficult to detect under such low magnifications. Using the Chem*Scan* system the magnification used during the validation procedure does not affect the speed of analysis consequently a very high magnification of x630 is routinely employed which allows visualisation of even very badly labelled organisms. Perhaps the only drawback of the ChemScan RDI system is that the analysis is performed upon membranes which do not permit oocyst confirmation using Nomarski differential interference microscopy (DIC). This technique necessitates the illumination of samples from below. However, the membranes used within this study do not allow the transmission of light in such a manner and consequently work is ongoing in this laboratory to identify membranes which would permit the use of DIC microscopy.

Although our findings indicated that the Chem*Scan* RDI system has the capability to detect all labelled oocysts upon a membrane surface not all fluorescent events detected by the instrument were confirmed as being *Cryptosporidium* oocysts. The interfering fluorescent events observed appeared to be particulate or cellular material to which the FITC-mAb had non specifically bound. Such findings have been reported previously in which monoclonal antibodies developed to be specific to *Cryptosporidium sp.* have been found to cross react with particulate material and algal/yeast cells[5]. Nevertheless, the number of such events detected by the system were found to be low and consequently did not significantly affect the analysis time.

In conclusion, our data suggests that IMS may be employed for the efficient isolation of oocysts from water concentrates. However, further work is required in order to increase the amount of sample which may be analysed using this assay.

During these studies a laser scanning instrument (ChemScan *RDI*) was employed for the rapid detection and enumeration of oocysts. This instrument was found to be extremely reliable, having the capability to detect all oocysts present upon a membrane surface. In addition the number of fluorescent events detected and their characteristics can be stored electronically. The Chem*Scan* RDI system therefore offers a reliable and rapid procedure for the detection and enumeration of oocysts in water samples and is a suitable alternative to direct epifluorescence microscopy.

References

1. Z. Bukhari, R.M. McCuin, C.R. Fricker, and J.L. Clancy, *Appl. Environ. Microbiol.*, 1998, **64**, 4495
2. P.A. Rochelle, R. DeLeon, A. Johnson, M.H. Stewart and R.L. Wolfe, *Appl. Environ. Microbiol.*, 1999, **65**, 841.
3. C.R. Fricker, *In* W.P. Betts, D.P. Casemore, C. R. Fricker, H.V. Smith and J. Watkins (ed.), 'Protozoan parasites and water'. The Royal Society of Chemistry, Cambridge, England, 1995, p91.
4. T.N. Whitmore, and M. Nazir, 'Improved techniques for the recovery of *Cryptosporidium parvum* from water'. *In* C.R.Fricker, J.L. Clancy and P.A. Rochelle (ed.)., International symposium on waterborne *Cryptosporidium*. American Water works Association, Denver, Colorado, p103.
5. M.R. Rodgers, D.J. Flanigan and W. Jakubowski, *Appl. Environ. Microbiol.*, 1995, **61**, 3759-3763.

POLYMERASE CHAIN REACTION FOR THE DETECTION OF PARASITES AND VIRUSES

C.P. Gerba, K.A. Reynolds, S.E. Dowd, and I.L. Pepper

Department of Soil, Water and Environmental Science
University of Arizona
Tucson, AZ 85721
USA

1 INTRODUCTION

The advent of molecular biology has created a new age of methodologies for the detection of pathogenic microorganisms in the environment. Techniques such as the polymerase chain reaction (PCR) have made possible very specific and sensitive methodologies for the detection of food and water-borne enteric viruses and protozoa. Traditional methods for the detection of these pathogens have been time consuming and costly. In addition, many pathogens cannot be grown in the laboratory, making their detection difficult without molecular techniques. PCR is especially useful for virus detection where cultivation techniques may require many weeks for the isolation and identification of an agent. While offering many advantages over current methodology, many limitations need to be overcome before the benefits can be fully realized. These include the presence of interfering or inhibiting substances, small assay volumes, the fact that the method is only semi-quantitative, and the inability of the method to assess infectivity.

1.1 Polymease Chain Reaction (PCR)

PCR is a technique used to amplify the amount of target DNA to 10^6 or more, allowing the detection of small numbers of microorganisms. PCR is a relatively simple enzymatic reaction that uses a DNA polymerase enzyme to repeatedly copy a target DNA sequence during a series of 25-30 cycles. During each cycle of PCR, the amount of target DNA is doubled resulting in an exponential increase in the amount of DNA. In theory, 25 cycles results in 2^{25} amplification, but in practice approximately a 10^6 fold increase in the amount of target present occurs since the efficiency of amplification is not perfect. The PCR product is typically visualized by agarose gel electrophoresis and size estimated by standards of known size.

A typical cycle of PCR has three steps. The first step involves the denaturation of double stranded DNA (ds DNA), that contains a target sequence, into two single strands of target or template DNA (ss DNA). Added to the reaction mixture are two different short pieces of single stranded DNA called primers that have been carefully chosen and commercially synthesized. Primers are oligonucleotides that have a complementary sequence to the target ss DNA template, so they can hybridize or anneal to this DNA defining the region of amplification. One primer is described as the upstream primer while the other is the downstream primer. This description simply defines the location of where the primers anneal on the ss DNA template based on sequence numbers. The second step in the PCR cycle is primer annealing which consists of the primers hybridizing to the appropriate target sequence. The third and final step is extension. Here, a DNA Polymease synthesizes a complementary strand to the original ss DNA by the addition of appropriate bases to the primers that have already hybridized to the target. The net result at the end of a cycle is two double-stranded molecules of DNA identical to the original double stranded molecule of DNA. Repeating the process results in PCR amplification of the DNA and an exponential increase or amplification of the copies of the original DNA present.

Key to the PCR cycle is that each of the three steps of PCR amplification occurs at different but defined temperatures for specific time interval. Generally, these three steps are repeated 25 to 30 times to obtain sufficient amplification of the target DNA. These cycles are conducted in an automated, self-contained temperature cycler or thermocycler. Thermocyclers are relatively inexpensive, but allow precise temperature control for each step, and provide for a substantial reduction in labor. The actual amplification takes place in small (approx. 0.2-0.5 ml) microfuge tubes and commercial kits are available that provide all the necessary nucleotides, enzymes and buffers required for the reaction.

Temperature is a critical part of the PCR process. Denaturation, of the target sequence or template, occurs at a temperature greater than the melting temperature of the DNA. For most PCR reactions this is standardized at 94°C for 1.5 min, because it guarantees complete denaturation of all DNA molecules. Primer annealing occurs at a lower temperature, typically 50 to 70°C for 1 minute, depending on the base composition of the primer. It is possible for a primer to anneal to a DNA sequence, which is similar to the correct target sequence, but which contains a few incorrect bases. This will result in the incorrect amplification of the DNA, which is termed nonspecific amplification, and gives a false positive result. The higher the temperature of annealing, the more specific the annealing is, and thus the extent of nonspecific amplification is reduced. However, as annealing temperature increases, PCR sensitivity normally decreases. The final step of PCR is extension. The essential component of this reaction is the polymerase enzyme, such as Taq polymerase which sequentially adds bases to the primers. This enzyme was obtained from the thermophilic bacterium *Thermus aquaticus*, and is uniquely suited for PCR because it is heat stable, withstanding temperatures up to 98°C, and can, therefore, be reused for many cycles. The extension step is normally 1 minute in length and is performed at 72°C. A 25-cycle PCR reaction may take about 3 hours, including ramp times between each step although the time varies depending on the type of thermocycler. Ramp time is simply the time interval it takes the thermocycler to go from one temperature to the next. It is important to keep in mind that for each primer pair, the research must initially optimize the conditions of temperature, incubation time and concentration of the various reaction components to obtain the desired results.

The choice of the primer sequences is critical for successful amplification of a specific DNA sequence. As in the case of gene probes, primer sequences can be deduced from known DNA sequences. The overall choice of primers is guided by the objectives of the investigator. If detection of a target DNA that is specific to a given species or genus of a virus is required, then only sequences unique to that virus or parasite are appropriate for the design of the primers. However, some objectives may require primers from sequences conserved across a group of viruses, for example, enteroviruses. Conserved sequences are those that are similar and are found in related viral species. Since the location of the primers within the genome defines the size of the amplification product, this theoretical size can be compared to the actual size of the product obtained on an electrophoresis gel containing DNA standards. Further confirmation that the amplified product is the DNA sequence of interest can be obtained by the use of a gene probe. Southern blot hybridizations are used to confirm that the PCR product is the gene target of interest by using a probe internal to the PCR product sequence. The probe will only hybridize to the complementary region of the PCR product if successful amplification of the correct sequence including internal regions has occurred.

1.2 Reverse Transcriptase - PCR

Most water and food borne viruses contain only RNA. Before PCR can be used to detect these viruses, a DNA copy of the RNA must first be accomplished. This involves the use of the enzyme reverse transcriptase and a method known as reverse transcriptase-PCR (RT-PCR). In RT-PCR, the first step is to make a DNA copy of the RNA sequence of interest. This copy is known as complementary DNA or cDNA The key enzyme in the reaction is reverse transcriptase which is a RNA-dependent DNA polymerase used to synthesize DNA from an RNA template. The first step in RT-PCR is performed using either the downstream antisense primer, which has the complementary sequence to the RNA, or random hexamers (small ~ 6 bp random primers) to make a complete complementary DNA (cDNA) copy of the RNA molecule. Thus, the cDNA molecule is the complement of the DNA strand. Normal PCR is then performed. During the first cycle, a complementary strand of DNA to the cDNA is constructed. Following this, PCR precedes normally from the double standard DNA template. A second useful application of RT-PCR is in the analysis of mRNA which allows estimation of metabolic activity. Since only viable protozoan parasites produce mRNA, it has been used to assess their viability in environmental samples.

2 VIRUS DETECTION BY PCR

Because of the importance of low concentrations of viruses in water, large volumes of water (usually 100-1000 liters) must first be processed to concentrate them. This is usually accomplished by their adsorption to positively charged microporous filters[3]. The viruses are then eluted from the filters and concentrated to a volume of 25-30 ml by a bioflucculation procedure involving the use of beef extract. Concentrated along with the viruses are various organic and inorganic contaminates which may interfere with PCR[4]. Viruses must also be extracted from food[5,6] before detection by PCR which also results in the presence of substances that interfere with PCR detection. Various approaches have been developed, such as, gel filtration, magnetic bead antibody capture[8], spin column chromatography, and ultrafiltration[9] to remove these substances, but significant loss of virus can occur during such processing (99% or more). In addition, to being time consuming and laborious in processing, some times no treatment appears capable of removing all of the contaminants in environmental samples.

Another drawback has been the small volume of sample which can be analyzed by PCR, usually only 10-50 µl. Typical concentrates from water and food are in the 20-30 ml volume range, allowing only a small fraction to be tested. Thus, a significant concentration of the virus must be present in the sample to allow detection. Concentrate volumes could be further reduced, but this requires additional processing with the additional problem of further concentration of PCR interfering substances.[10] Even given these limitation, PCR has

Table 1   *Limitations with the use of PCR for Virus and Parasite Detection in Environmental Samples*

• Small assay volumes (typical samples must be further concentrated; immunomagnetic separation techniques have proven successful)
• Inhibition by interfering substances (samples must be pretreated)
• Inability to determine live from dead organisms (can be combined with cell culture or potentially messenger RNA detection in parasites) [8, 12]
• Nonquantitative assay (may be used with an MPN approach)[13]

been useful in the detection of viruses in groundwater[4,11], wastewater, food[5,6,7], and sewage sludge[12]. A recent survey of wells, used as drinking water, in the United States found 30.1% of 133 wells were found to be contaminated with enteroviruses[11] using PCR for their detection.

3 VIABILITY ASSESSMENT OF ENTERIC VIRUSES BY INTEGRATED CELL CULTURE PCR (ICC-PCR)

A new procedure identified as integrated cell culture - PCR (ICC PCR) was recently introduced by Reynolds et al[12]. Here, cell culture monolayers in flasks are inoculated with pure cultures of virus, or aliquots of environmental samples, and the flasks incubated for 2-3 days. Following incubation, flasks are frozen, resulting in cell lysis and subsequent release of virus particles. RT PCR is then conducted on the cell culture lysate. In essence, this is a biological amplification followed by an enzymatic amplification, and integrated together, the procedure can identify virus in 3 days as opposed to the 10-15 days required for cell culture alone. Thus, this new technique greatly enhances the speed of virus detection. It has other important advantages as well. Since growth of the virus precedes PCR amplification, only infectious viruses are detected. In addition, sensitivity of ICC PCR is better than direct RT PCR since larger sample volumes can be added to the initial cell culture. Dilution of sample, within the cell culture, also dilutes any PCR inhibitory substances that may be present in the environmental sample, although cell culture components can also interfere. Finally, since ICC PCR positive samples are confirmed by PCR, there is no need to run an additional cell culture assay to confirm positives, as in the case when the virus is detected by cytopathogenic effects (CPE) or plaque forming unit (PFU) methods (Table 2).

Our recent application of ICC-PCR to the study of enteroviruses in heavily disinfected wastewater and groundwater has revealed the occurrence of non-cytopathogenic (CPE) producing enteroviruses. These enteroviruses do not produce CPE even after 3 to 4 passages in cell culture. They probably represent varients of enterovirus strains which do not produce CPE in cell culture. Studies have shown that for every plaque forming unit of poliovirus, observed in cell culture, 100 intact virus particles can be observed under the electron microscope[15] and in non-cell culture grown strains (direct detection from feces), this may be as great as 1:50,000[16]. These additional virions may represent non-CPE varients which are only detectable by ICC-PCR. Thus, another significant advantage of ICC PCR is its ability to detect non-CPE producing viruses, which may occur in greater numbers than CPE producing viruses.

Table 2 *Comparison of Methods for the Detection of Enteric Viruses in Water Concentrates*

Advantages	Direct PCR	Cell Culture	ICC/PCR
Sensitive	yes	yes	yes
Specific	yes	no	yes
Rapid	yes	no	yes
Able to examine large equivalent volumes	no	yes	no
Minimizes inhibition/toxicity	no	no	yes
Detects noncytopathogenic viruses	yes	no	yes
Detects infectious viruses only	no	yes	yes

4 DETECTION OF PROTOZOAN PARASITES IN WATER

The ability of *Giardia intestinalis* and *Cryptosporidium paruum* to cause waterborne disease is well documented. However, current analytical techniques are difficult and do not allow an assessment of viability. While a number of PCR methods have been described for detection of both *Giardia* and *Cryptosporidium*[1,4,8], they suffer from some of the same limitations as PCR detection of enteric viruses i.e. sensitivity and the presence of interfering substances. One approach that appears to overcome some of these difficulties is the use of specific antibody coated magnetic beads. Because mRNA has a short existence in dead cysts or oocysts RT-PCR can also be used to assess viability of the organisms detected. Studies with field samples indicate that RT-PCR is as sensitive or more sensitive than current immunofluorescence (IF) methods[1].

Another application of PCR is for the detection of protozoan parasites which can not be detected or confirmed by other techniques, such as IF. An example of this is the microsporidia. "Microsporidia" is a nontaxonomic name used to describe protozoan parasites belonging to the phylum *Microspora* of which there are more than a 1,000 species infecting a wide variety of animals and insects[16]. Only a few are known to cause infections in humans. Two species in particular, *Enterocytozoan bieneusi* and *Encephalitozoon intestinalis* have been shown to be responsible for intestinal illness in AIDS patients[17]. Specific antibodies have not been developed for the detection of these organisms in environment, non-specific activity continues to be a problem. To overcome these difficulties, Dowd et al[18] used community DNA extraction followed by microsporidium-specific PCR amplification. Following this, product DNA sequencing followed by computer based DNA homology analysis allowed for specific identification of the organisms in water. Such approaches in the future can be used to identify almost any suspected waterborne protozoan parasite in water or food.

5 SUMMARY

PCR in combination with product sequencing offers the potential for the detection and identification of almost any pathogenic virus or parasite in water. For its full potential to be realized, a number of limitations still need to be overcome (Table 1), however, these do not appear insurmountable given the recent advances in this technology. This technology has also revealed that we may have only been studying a small fraction of all the enteric viruses that are present in contaminated water (e.g. non-CPE producing viruses). PCR still holds the promise of a more rapid, less costly, and sensitive methodology for detection of these pathogens in the environment.

References

1.   C. Kaucner and T. Stinear, *Appl. Environ. Microbial.*, 1998, **64**, 1743.
2.   M. Abbaszadegan, M. S. Humber, C. P. Gerba, and I. L. Pepper, *Appl. Environ. Microbiol.*, 1997, **63**, 324.
3.   C. P. Gerba, and S. M. Goyal, 1992, 'Methods in Environmental Virology,' John Wiley and Sons, New York.
4.   M. Abbaszadegan, M. S. Huber, C. P. Gerba, and I. L. Pepper, *Appl. Environ. Microbiol.*, 1993, **59**, 1318.
5.   R. L. Atmar, F. H. Neill, C. M. Woodley, R. Manger, G. S. Fout, W. Burkhardt, L. Leja, E. R. McGovern, F. Le Guyader, T. G. Metcalf, and M. K. Estes, *Appl. Environ. Microbiol.*, 1996, **62**, 254.
6.   M. Y. Deng, S. P. Day, and D. O. Cliver, *Appl. Environ. Microbiol.*, 1994, **60**, 1927.
7.   O. Traore, C. Arnal, B. Mignotte, A. Maul, H. Laveran, S. Billaudel, and L. Schwartzbrod, 1998, **64**, 3118.
8.   D. W. Johnson, N. J. Pienlazek, D. W. Griffin, L. Misener, and J. B. Rose, *Appl. Environ. Microbiol.*, 1995, **61**, 3849.
9.   K. J. Schwab, R. DeLeon, and M. D. Sobsey, *Appl. Environ. Microbiol.*, 1995, **61**, 531.
10.   J. F. Ma, C. P. Gerba, and I. L. Pepper, *J. Virological Methods*, 1995, **55**, 295.
11.   M. Abbaszadegan, P. Stewart, and M. LeChevallir, *Appl. Environ. Microbiol.*, 1999, **65**, 444.
12.   K. A. Reynolds, C. P. Gerba, and I. L. Pepper, *Appl. Environ. Microbiol.*, 1996, 62, 1424.
13.   J. B. Rose, X. Zhou, D. W. Griffin, and J. H. Paul, *Appl. Environ. Microbiol.*, 1997, **63**, 4564.
14.   T. M. Straub, I. L. Pepper, M. Abbaszadegan, C. P. Gerba, *Appl. Environ. Microbiol.*, 1994, **60**, 1014.
15.   D. G. Sharp, Transmission of Viruses by the Water Route, Berg, G., ed., John Wiley, New York, p.193.
16.   R. L. Ward, D. R. Knowlton, and M. J. Perce, *J. Clin. Microbiol.*, 1984, **19**, 748.
17.   A. Curry, and E. U. Canning, Human microsporidiosis, *J. Infect.*, 1993, **27**, 229.
18.   S. E. Dowd, C. P. Gerba, and I. L. Pepper, *Appl. Environ. Microbiol.*, 1998, **64**, 3332.

A RAPID DETECTION METHOD AND INFECTIVITY ASSAY FOR WATER-BORNE *CRYPTOSPORIDIUM* USING PCR AND IN-VITRO CELL CULTURE

R. De Leon, P. A. Rochelle, H. Baribeau, M. H. Stewart and R. L. Wolfe

Metropolitan Water District of Southern California
Water Quality Laboratory, La Verne, California   91750

1 INTRODUCTION

Recent outbreaks of cryptosporidiosis associated with drinking water and boil water advisories affecting hundreds of thousands or millions of people have ensured that waterborne *Cryptosporidium parvum* continues to be a significant concern facing the water industry and public health agencies. The organism's relative resistance to chlorine disinfection at the doses typically applied in water treatment plants, the lack of effective chemotherapeutic agents, and the potentially fatal consequences for immune compromised individuals who become infected makes waterborne *C. parvum* a critical issue. However, the widespread occurrence of *C. parvum* oocysts in raw water and recent episodes of oocyst contaminated drinking water, without concomitant increases in reported cases of cryptosporidiosis, raise questions regarding the infectivity of waterborne *C. parvum* oocysts.

The standard method used in the United States for detection of *Cryptosporidium* oocysts in water (filtration, centrifugation, and detection using an immunofluorescent assay [IFA] and microscopy) demonstrates poor and variable recovery efficiencies (0 to 140%), lacks specificity, is labor intensive, and can not determine whether oocysts are viable or infectious[1]. Consequently, there is a need for methods which provide rapid, sensitive and reliable oocyst detection. Also, methods are required to determine the infectivity of oocysts, which will allow water utilities to more fully assess the public health significance of *C. parvum* in water. Moreover, unlike the food industry and clinical laboratories, water quality laboratories lack simple and rapid methods for pathogen detection that are readily amenable to automation. Therefore, the objectives of this study were to: (i) develop a simple, rapid, robust, and sensitive detection method for waterborne *C. parvum*, using PCR-based technology with the potential for automation; and (ii) develop an in-vitro cell culture-based infectivity assay for waterborne oocysts.

2 RAPID DETECTION METHOD

Polymerase chain reaction (PCR)-based detection assays offer the advantages of high sensitivity, absolute specificity (depending on primer design and selection), rapid sample processing, detection of multiple pathogens simultaneously, and the ability to screen many samples concurrently (up to 384 with some thermal cyclers). At least 10 sets of PCR primers have been described for the detection of *C. parvum* in a variety of types of sample and many have been evaluated and compared for the detection of *C. parvum* oocysts in water[2,3]. Following evaluation of published primers, we designed *C. parvum*-specific PCR primers which amplified a 361-bp fragment from a heat shock protein gene (*hsp*70)[4]. The primers detected less than 10 oocysts seeded into 65 to 100 liters of concentrated environmental water samples with turbidities ranging from 0.4 to 12.5 NTU, and were compatible with multiplex PCR for detection of *C. parvum* and *Giardia lamblia*[5]. The DNA extracted from 61% of environmental samples (n=41) inhibited PCR without prior purification by spin column chromatography.

This PCR-based approach using the *hsp*70 primers was adapted to a relatively simple, semi-automated, 96-well microplate format to provide rapid detection with high sample throughput. The *hsp*70 fragments were simultaneously amplified and labeled with digoxigenin (DIG) by PCR. The DIG-labeled amplicons were then captured by an internal oligonucleotide probe which was attached to the wells of a microplate via a biotin-streptavidin bond. The immobilized amplicons were detected using an anti-DIG antibody conjugated to alkaline phospatase (AP) and an AP sensitive colorimetric substrate which produced a purple precipitate in positive wells. The intensity of the color reaction was proportional to the number of oocysts detected. Colored reactions products were detected either visually or using an automated microplate reader. This detection procedure eliminated the subjective identification of positive samples which is a major limitation of the microscope-based IFA method. The entire PCR and liquid hybridization procedure was conducted in approximately 6 hours which significantly decreased the analysis time compared to PCR followed by membrane hybridization (>20 hours) or the current IFA method (1-2 days). A major advantage of this technique is that it has the potential to analyze up to 384 samples simultaneously without greatly increasing the processing time (thermal cyclers and microplate readers which can handle 384 samples are available).

3 INFECTIVITY ASSAY

The three techniques available to determine pathogen infectivity are human volunteers, animal models, and in vitro cell culture. Mouse infectivity assays are currently the accepted technique for measuring *C. parvum* infectivity but this technique raises ethical issues and is too costly and impractical for routine monitoring use by the water industry. Human volunteer studies are also not practical. Cell culture provides a promising and practical alternative technique for determining infectivity.

An in-vitro infectivity assay was developed using monolayers of human cells grown on adapted microscope slides[5]. Infections were detected by reverse transcriptase (RT)-PCR targeting *hsp*70 mRNA using the same primers described above. Extensive control infections demonstrated that inoculum oocysts were not detected by this method. Negative control infections that did not produce amplification products included uninfected cells, cells infected with heat and formalin inactivated oocysts, infections with unseeded source water, and inoculation of cells that did not support infection of *C. parvum*. Only infectious oocysts invaded the human cells and produced various stages of the *C. parvum* life cycle which actively transcribed *hsp*70 mRNA. The sensitivity of this assay, combined with immunomagnetic purification of oocysts, was 10 infectious oocysts or less. The *hsp*70 gene is an ideal target for RT-PCR detection of infections because, under certain conditions, parasites contain large quantities of *hsp* mRNA and it has been suggested that the heat shock response plays a major role during host invasion by parasites[6].

Different cell lines were evaluated for their ability to support in-vitro growth of *C. parvum* and protocols were established which allowed routine infection of cell cultures. Oocysts were recovered from seeded environmental water samples using USEPA Method 1622 involving capsule filtration and immunomagnetic separation (IMS) of oocysts[7]. Studies in this laboratory have demonstrated 64 to 100% recovery of low concentrations of oocysts seeded into environmental water concentrates by IMS alone[8] and average recovery efficiencies of 70% for the entire Method 1622 procedure. The purified oocysts were inoculated onto monolayers of Caco-2 or HCT-8 cells and incubated at 37°C in slide chambers for 24-72 h. Infectious oocysts were detected by *C. parvum*-specific RT-PCR of mRNA purified from total extracted RNA. RT-PCR targeting *C. parvum*-specific *hsp*70 mRNA demonstrated detection of less than 10 infectious oocysts.

This infectivity assay was used to demonstrate that the infectivity of oocysts recovered using Method 1622 was not impaired. Also, purification of oocysts by IMS prior to infection significantly increased the sensitivity of the infectivity assay. The assay was used to demonstrate the in-vitro infectivity of a variety of fresh animal isolates but considerable variability in the infectivity of a single isolate of *C. parvum* was noted for oocysts prepared at different times from different source animals. For example, 10,000 oocysts of the IOWA isolate always produced infection but detectable infection was only obtained 60% of the time with 1,000 oocysts, even though the same isolate was used throughout. The reasons for this variability are currently under investigation. The infectivity assay was also used to demonstrate the effectiveness of UV irradiation for *C. parvum* inactivation.

To improve the robustness of the assay, a variety of procedures for extraction of mRNA from infected cell cultures were evaluated. Total RNA extraction using a S.N.A.P. kit (Invitrogen) which incorporated a DNase digestion step, followed by mRNA purification using oligo (dT)$_{16}$ cellulose columns (Invitrogen) was found to be the most effective procedure. Other procedures that were evaluated included TriReagent (Sigma) for total RNA extraction, and oligo (dT)$_{16}$ paramagnetic beads (Dynal) and magnetic columns (Miltenyi) for purification of mRNA. Although these methods did produce high yields of mRNA, it was usually contaminated with trace amounts of DNA. Such DNA contamination would leave to false-positive infectivity results since non-infectious oocysts inoculated onto monolayers may contain DNA which would be detected by the amplification reaction.

A disadvantage of the RT-PCR detection approach is that it can only be considered semi-quantitative. Although it can be used to demonstrate relative differences in infectivity, it can not be used for absolute quantitation. Therefore, an in-situ hybridization (ISH) procedure was developed using DIG-labeled probes targeting the same *C. parvum*-specific *hsp*70 sequence described above. Probes were hybridized directly to specific nucleic acids in *C. parvum* infectious stages in the cell monolayers. Localized detection of positive hybrids was achieved using anti-DIG-AP and the same colorimetric assay described for the rapid detection method. Signals were enumerated visually by simple light microscopy. The advantage of the ISH procedure is that it allows for direct quantitation of infectious foci in the cell monolayers while retaining the high specificity of nucleic acid probes. Maximum sensitivity was obtained using full length probes with multiple DIG labels but shorter oligonucleotide probes provided greater specificity. The sensitivity of single DIG-labeled oligonucleotide probes was increased by the application of signal amplification technology.

4 CONCLUSIONS

The combination of molecular detection methods and in-vitro cell culture provides the water industry with a tool to more fully assess the public health significance of waterborne *Cryptosporidium*. With the recent advent of extremely rapid thermal cyclers and automated detection methods, PCR-based techniques can reduce oocyst detection time from days to a few hours. A relatively simple and robust infectivity assay for *C. parvum* will allow water utilities to measure the efficacy of disinfection practices and monitor the occurrence and persistence of infectious oocysts in environmental waters. The commercial availability of ready to use kits and products for the various stages in the procedure, and the improvements and simplifications we have made to the method should ensure the assay is applicable to a wide range of water quality laboratories.

References

1.   W. Jakubowski, S. Boutros, W. Faber, R. Fayer, W. Ghiorse, M. LeChevallier, J. B. Rose, S. Schaub, A. Singh and M. H. Stewart. *J. Am. Water Works Assoc.*, 1996, **88**,107.
2.   P. A. Rochelle, R. De Leon, M. H. Stewart and R. L. Wolfe. *Appl. Environ. Microbiol.* 1997, **63**, 106.
3.   S. D. Sluter, S. Tzipori and G. Widmer. *Appl. Microbiol. Biotechnol.* 1997, **48**, 325.
4.   P. A. Rochelle, D. M. Ferguson, T. J. Handojo, R. De Leon, M. H. Stewart and R. L. Wolfe. *J. Euk. Microbiol.*, **43**, 72S.
5.   P. A. Rochelle, D. M. Ferguson, T. J. Handojo, R. De Leon, M. H. Stewart and R. L. Wolfe. *Appl. Environ. Microbiol.* 1997, **63**, 2029.
6.   B. Maresca and L. Carratu. *Parasitol. Today*, **8**, 620.
7.   United States Environmental Protection Agency. Method 1622: *Cryptosporidium* in water by filtration/IMS/FA. Jan. 1999. Office of Water, Washington, DC.
8.   P. A. Rochelle, R. De Leon, A. Johnson, M. H. Stewart and R. L. Wolfe. *Appl. Environ. Microbiol.* 1997, **65**, 841.

Acknowledgments

This work was partly supported by research grants from the U.S. Environmental Protection Agency (R825146-01-0) and the American Water Works Association Research Foundation (PFA 2565).

MOLECULAR TECHNIQUES FOR THE DETECTION OF BACTERIAL PATHOGENS IN DRINKING WATER

Nigel Lightfoot,[1] Martin Pearce,[2] Barbara Place[1] and Chaminda Salgado[2]

[1] Public Health Laboratory Service North, 17-21 Dean Street, Newcastle upon Tyne NE1 1PQ
[2] Chemical and Biological Defence Sector (CBD), DERA Porton Down, Salisbury, Wiltshire SP4 0JQ

1 INTRODUCTION

There will occasionally be the requirement, for example in developing contamination situations, to detect bacterial pathogens in addition to indicator organisms in drinking water. The established methods are based on the concentration of the micro-organisms by membrane filtration and different selective cultural techniques to isolate the organisms prior to biochemical and serological confirmation. These procedures are however time consuming and usually take up to four or five days. There is a definite need for rapid methods. Molecular techniques to detect bacterial DNA now offer an opportunity for the detection of bacterial pathogens in a few hours rather than isolation over several days. Our aim was to set up a primarily qualitative test to detect the presence of salmonellae, campylobacters and toxigenic strains of *Escherichia coli* O157 in samples of drinking water of volumes 100 – 1000 ml using the polymerase chain reaction (PCR) technique. Detection of target bacterial DNA sequences does not give any information about the viability of the organisms. Viability can be determined using traditional culture methods performed in parallel.

2 SELECTION OF PCR PRIMERS

Primers were chosen which were specific for the target organisms and the selection of the target gene sequences would ideally give an indication of their pathogenicity status.

For detection of organisms of the genus *Salmonella*, primers were based on the sequences of the salmonella *inv*E and *inv*A genes whose protein products are necessary for invasion of epithelial cells[1]. Amplification of a product which overlapped the junction between the *inv*E and *inv*A genes eliminated the occurrence of false positives as seen with the *inv*A gene only and resulted in 457bp product.

For the detection of the thermophilic enteropathogenic campylobacters commonly associated with human infections, namely *C. jejuni*, *C. coli* and *C. upsaliensis,* the primers described by Jackson *et al* were chosen[2]. These primers amplify a 256 bp product and are designed from the sequence of an open reading frame thought to be a haem-copper oxidase domain adjacent to and downstream from a novel *C. jejuni* two-component regulator gene.

The approach to detecting *E. coli* O157 and other verocytotoxic *E. coli* was to use the primer pairs described by Pollard *et al.* for the detection of verotoxin 1 and 2 genes (i.e. Shiga-like toxin I and II genes)[3]. The VT1 primers amplify a 130 bp fragment of the gene coding for the B subunit of the toxin, and the VT 2 primers amplify a 346 bp fragment in the region coding for the A subunit.

3 MATERIALS AND METHODS

Before testing seeded water samples, each PCR was tested against pure cultures of target and non-target organisms. The Campylobacter PCR has been tested against a number of laboratory isolates of *C. jejuni, C. coli* and *C. upsaliensis* and 27 isolates of Gram negative bacteria from 16 genera and 6 Gram positive organisms from 5 genera[2]. The Salmonella PCR was tested with 47 salmonellae from 32 serovars and 53 organisms from 28 genera including *Citrobacter freundi* which has been described as causing false positives. The *E. coli* VT1/2 PCR was tested against a range of isolates positive for both verocytotoxin 1 (VT1) and verocytotoxin 2 (VT2), VT1 alone, VT2 alone and non-VT producing *E. coli* O157. Non-target organisms tested were 18 Gram-negative organisms from 6 genera.

The details of the PCR conditions for each organism are shown in Tables 1 and 2. The PCR assays were performed using a Perkin Elmer 2400 thermal cycler.

The products were visualised by ultraviolet illumination of ethidium bromide-stained products after gel electrophoresis using 1.5% (w/v) agarose (Pharmacia) and 0.5 mg l^{-1} ethidium bromide run at 90 volts for 1 hour.

Having established the working PCR methods using pure cultures of organisms the sensitivity of the assay to detect organisms in seeded drinking water samples was measured. The seeded samples were prepared by diluting an overnight culture of each organism, 1 ml in 100 ml water and a further six ten-fold dilutions 10 ml + 90 ml. These seeded samples were examined by PCR and viable counts determined by plate culture. The seeded water samples were concentrated by membrane filtration using several different types of membranes. The samples of volumes 100 ml – 500 ml were applied to 25 mm diameter 0.45 μm pore size membranes in a reusable and sterilisable Swinnex membrane filter holder (Millipore), for Campylobacter a 0.22μm pore size membrane was used. The DNA was extracted from the membrane using 1.0 ml DNA extraction reagent (Chelex) in a sterile 1.5 ml Eppendorf tube by vortexing and heating to 98°C for 10 minutes and concentrated by centrifuging in a Millipore Ultrafree CL filter at 1620g for 3 minutes. An aliquot of the sample was taken prior to the concentration step as potential PCR inhibitors may also be concentrated in this step. A comprehensive control strategy has been developed to remove as far as possible false positive and false negative results from assays the results of which can have far reaching consequences (Figure 1).

Controls 1 and 2 were positive and negative filtration/whole process controls. The negative control for all assays was uninoculated sterile distilled water and was to control for contamination. Positive controls consisted of 100 ml sterile distilled waters innoculated with a lenticule containing 10^4 organisms of an NCTC strain of the target organism. Lenticules have been developed in the Newcastle laboratory, they are lentil sized and contain defined numbers, within Poisson distribution, of organisms that have been control dried[4].

Positive and negative controls 3,4, were also used at the PCR stage. The negative PCR control consisted of uninnoculated chelex reagent aliquots which had been subjected to the DNA extraction process in advance and frozen ready for use. The positive PCR control

Table 1 *Master Mix*

Reagents	*Salmonella* spp.		*Campylobacter* spp.		*E. coli* O157	
	μl	Final conc.	μl	Final conc.	μl	Final conc.
H₂0	30.5		27.1		27.3	
Perkin Elmer BII x 10	5	x 1	5	x1	5	x 1
PE 25mM MgCl₂	3	1.5 mM	6	3 mM	6	3.0 mM
4 x PE 10mM dNTPs (1 each A,T,G,C)	4	200 μM each	4	200 μM each	1.5	75 μM each
PE TAQ 5U/μl	0.5	2.5 U	0.4	2 U	0.2	1.0 U
Primers	2 x 1.0	1.0 μM each	2 x 1.25	0.5 μM	4 x 1.0	1.03 μM 2.05 μM 3.05 μM 4.03 μM
Master mix volume	45		45		44	
DNA template	5		5		5	
Total volume	50		50		49*	
PRIMERS	*inv* A/E invasion of epithelial cells		ORF-C Downstream from Novel *C. jejuni* two-component regulator gene		VT 1 VT 2 Verotoxin Genes	
PRODUCT SIZE	457 bp		256 bp		130 bp 346 bp	

*NB: Plus 1 μl Internal Control/H₂0 = 50 μl**

Table 2 *Thermal - Cycler*

CYCLE	UNIVERSAL CYCLE	*Salmonella* spp.	Thermophilic *Campylobacter* spp.	*E. coli* O157
Denature template	94°C 2 min	94°C 1 min	94°C 2 min	94°C 2 min
Denature	94°C 15 sec	94°C 15 sec	94°C 25 sec	94°C 30 sec
Anneal	52°C 15 sec x 35	52°C 15 sec x 35	55°C 40 sec x 40	60°C 30 sec x 35
Extension	72°C 30 sec	72°C 15 sec	72°C 60 sec	72°C 30 sec
Hold	72°C 2.5 min	72°C 2.5 min	72°C 5 min	72°C 4 min
Total time	1 hr 30 min	1 hr	2 hrs 20 min	1 hr 40 min

NB: The universal cycle can be used for *Salmonella* and *Campylobacter* spp.

Figure 1 *Processing of Water Samples for Direct Detection of Enteric Pathogens by PCR*

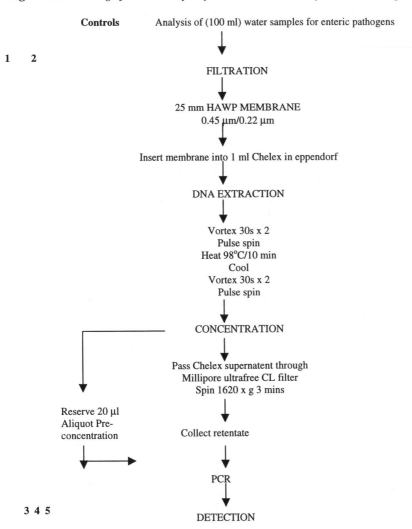

By gel electrophoresis (0.5 mg l^{-1} ethidium bromide)

consisted of template DNA prepared from plate cultures of the appropriate NCTC strain. Serial ten-fold dilutions of the DNA extracts were made to obtain a concentration close to the limit of detection. DNA at that dilution was aliquoted and stored ready for use at − 20°C. In addition an inhibition control, 5, was included for each test sample; an additional PCR for each post concentration test sample was ran in duplicate but contains a aliquot of the positive PCR control to detect inhibition. If inhibitors are present in the test sample they are most likely to be significant in the post concentration sample and therefore most easily detectable by this approach.

We have also developed an internal control template for the the the *E. coli* O157 assay. This template uses one of the VT1 primers and one of the VT 2 primers. The molecular weight of the product (218 bp) is intermediate between the two target products.

4 RESULTS

The membrane chosen to give the highest sensitivity was a nitrocellulose membrane (HAWP/GSWP, Millipore) of 25 mm diameter and with pore sizes of 0.22 μm for *Camypylobacter* species and 0.45 μm for *Salmonella* species and *E. coli* verocytotoxin 1 and 2. The PCR assays designed were confirmed as specific for the target organisms after testing each one against a wide range of non target organisms.

The sensitivity of the PCR assays as tested using seeded samples in internal quality control runs has been found to be as low as 10 organisms in each PCR tube. When processing simulated water samples of 100 ml through filtration, DNA extraction, PCR, and detection the limits of detection were found to be:

Salmonella sp.	10^3 orgs per 100 ml
Campylobacter sp.	10^3 orgs per 100 ml
E. coli O157 VT1	10^3 orgs per 100 ml
VT2	10^6 orgs per 100 ml

It was found that the internal control in the *E. coli* VT1/2 PCR assay competed with the two targets. Certain water sources, especially those containing humic acids, cause inhibition of the PCR reaction and it is essential to run the controls as described for each set of tests.

5 DISCUSSION

The aim has been to produce a truly rapid method producing an answer in 4-6 hours. The primers chosen have provided specificity but a further increase in sensitivity would be desireable. The methods must therefore be carefully controlled with strict adherence to details to minimise any loss of organisms or DNA during the process. The choice of membrane is important and we have found nitrocellulose (Millipore HAWP/GSWP) membranes statisfactory. Serveral other membranes that we have tested led to lower sensititives or even complete lack of detection at high seeding levels. The DNA extraction process is also significant: freeze/thaw methods and procedures involving multiple steps may give better recoveries of DNA than the Chelex extraction reagent and heating to 98°C for 10 minutes but have the disadvantage of taking considerably longer. The theoretical sensitivity of detection of a single organism was not achievable with real samples even though our methods were able to detect 10 organisms in a PCR tube. We have been able to increase the sensitivity of our method, per 100 ml , ten-fold by filtering 1000 ml samples but we are still working to further optimise the procedure at each stage. In addition, we are looking at short pre-incubation periods during the sample transport phase to further improve the sensitivity.

In practical terms we have established stringently controlled methods that are capable of detecting 10^3 organisms in one litre of water within six hours of the receipt of the sample. These rapid molecular methods are backed up by traditional culture methods but these can take up to 4 days.

We gratefully acknowledge the Drinking Water Inspectorate who supported this work with grant funding.

References

1   G.G. Stone, R.D. Oberst, M.P. Hays, S. McVey and M.M. Chengappa, *J. Clin. Microbiol.*, 1994, **32,** 1742.
2   C.J. Jackson, A.J., Fox, and D.M.Jones, *J. Clin Microbiol.*, 1996, **81**, 467.
3   D.R. Pollard, W.M. Johnson, H. Lior, S.D. Tyler and K.R Rozee, *J. Clin. Microbiol.*, 1990, **28,** 540.
4   A.A. Codd, I.R. Richardson and N. Andrews, *J. Appl. Microbiol.*, 1998, **85,** 913.

APPLICATION OF PCR FOR THE DETECTION OF VIABLE PATHOGENIC BACTERIA FROM WATER SAMPLES

Karine Delabre*, Marie-Renée de Roubin*, Véronique Lahoussine**, Paulina Cervantes*, and Jean-Claude Joret*.
* ANJOU RECHERCHE, Laboratoire Central Générale des Eaux, 1, Place de Turenne, 94417 Saint-Maurice cedex, FRANCE.
** AGENCE DE L'EAU SEINE NORMANDIE, 51, rue Salvador Allende, 92027 Nanterre cedex, FRANCE.

1 INTRODUCTION

The detection of pathogenic bacteria from water samples using traditional culture techniques is tedious and too long. Several analytical biaises can be introduced using these techniques. For example, the selective culture media used can inhibit the growth of the targeted bacteria because their composition is a compromise between factors able to inhibit the interfering flora and factors able to promote the growth of stressed bacteria. On the other hand, the techniques, like biochemical identification tests, necessary to identify the bacteria after isolation are sometimes inefficient because the expression level of enzymes can be very low in the case of stressed bacteria. All these disadvantages lead to sensitivity and specificity problems.

Molecular techniques, like PCR[1] (Polymerase Chain Reaction), can improve the detection of pathogenic bacteria from water samples by increasing the specificity and sensitivity of the detection. PCR can be used at different levels of the standard culture protocol. For example, PCR can be used as an identification method in place of biochemical identification tests. This approach allows to avoid problems linked with the biochemical tests. PCR can also be used as a direct detection method without selective culture steps. In this case, the utilization of both selective culture media and biochemical tests is unnecessary. However, when PCR is used as a direct detection method, the viability of detected bacteria can not be directly determined. PCR does not allow the differentiation between viable and dead bacteria[2]. It is necessary to associate with PCR an indirect approach for assessing the viability of PCR detected bacteria.

The aim of this study was to evaluate the PCR technology as an identification method for typical colonies of *Salmonella* and as a direct detection method for *Salmonella* bacteria from water samples. The different approaches developed were compared with the standard culture protocol and between them. The PCR step was developed for the detection of two other bacteria identified at the species level (*Aeromonas hydrophila*) or at the sub-species level (enterohemorragic *Escherichia coli*), and an indirect approach allowing to assess the viability of the PCR detected bacteria was developed. We can now use PCR as a direct detection method for three viable pathogenic bacteria from the same water sample.

2 MATERIALS AND METHODS

2.1 Bacterial strains : *Aeromonas hydrophila* 76.14, *Escherichia coli* 103571 (EHEC), and *Salmonella typhimurium* WG49 were obtained from the Institut Pasteur (Paris - France).

2.2 Water sample treatments : Water samples are concentrated by filtration through 0.45 μm pore size nitrocellulose filters (Sartorius - France). The filters are then vortexed in peptone broth (Sanofi-Pasteur - France) for recovering the bacteria. After removing the filters, the bacteria are either centrifuged at 4500 g for 20 min (without preenrichment) or cultivated at 37°C for 20 hours and are then centrifuged (with preenrichment).

2.3 Detection of *Salmonella* using culture techniques : after filtration of the water sample and a preenrichment step, the bacteria are recoved by centrifugation at 3000 g for 20 min. Afterwards, they are incubated in Rappaport-Vassiliadis broth (Merck) during 24 hours at 42°C. The broth is centrifuged at 3000 g during 20 min and the bacterial pellet is resuspended in 5 ml of broth. 100μl of this bacterial suspension are streaked on Rambach agar medium (Sanofi Diagnostic Pasteur - France). The typical colonies are subcultured on nutritive agar medium (Sanofi Diagnostic Pasteur - France), and then they are identified using API32E test (Sanofi Diagnostic Pasteur - France) for biochemical identification.

2.4 Confirmation of *Salmonella* colonies using PCR : the typical colonies subcultured on nutritive agar medium were resuspended in 100 μl of apyrogenic water and the bacterial DNA is released by heating at 94°C during 15 min. The PCR is performed in a total volume of 100 μl : 80 μl of reaction mixture and 20 μl of bacterial DNA. The composition of the reaction mixture and the reaction conditions were as previously described[3], except that BSA (Roche Molecular Diagnostics) (800 ng/μl) is added in the reaction mixture.

2.5 DNA purification and direct detection of pathogenic bacteria using PCR : these steps are performed as previously described[4].

3 RESULTS AND DISCUSSION

3.1 Use of PCR as an identification method for typical colonies of *Salmonella*

In order to introduce the PCR in our laboratory, *Salmonella* was chosen as a bacterial model for the following reasons : (i) the importance of this genus in terms of public health and, (ii) the existence of draft of a standard protocol for the enumeration of *Salmonella* using culture techniques. This protocol relies on several steps of non-selective and selective cultures. 47 river water samples were analyzed with this culture protocol and the typical colonies of *Salmonella* were confirmed using biochemical identification tests and PCR. 558 colonies were tested (table 1).

The results show that PCR and biochemical identification reached the same result in 84% of cases. In 9.7% of cases, the PCR results were negative while biochemical results were uncertain and could not discriminate between *Escherichia coli* and *Salmonella*. Two hypotheses can account for these results : (i) the bacteria belong to the *Salmonella* genus. In this case, the PCR negative results could be explained by the absence of detection of the two non mobile *Salmonella* species by PCR with the primers used, (ii) the bacteria belong to the *Escherichia coli* species. In this case, PCR and biochemical methods reach concordant results. Finally, in 6.3% of cases, the PCR results are positive and the biochemical results are negative or uncertain (absence of identification). These results show that PCR identification is more specific than biochemical identification. Furthermore, the utilisation of PCR allows a decrease in the delay of analysis.

Biochemical identification	PCR identification	Number of colonies	Percentages (%)
positive	positive	415	74.4
negative	negative	54	9.7
uncertain*	negative	54	9.7
negative	positive	10	1.8
uncertain	positive	25	4.4

Table 1 *Comparison between biochemical and PCR identification of presumed Salmonella colonies. * Salmonella spp or Escherichia coli.*

3.2 Use of PCR as a direct detection method

We have developed a PCR direct detection protocol for *Salmonella* after a non-selective pre-enrichment step[4]. This protocol relies on a partial DNA purification step followed by a nested-PCR detection in the presence of bovine serum albumine. 25 water samples (river, clarified, sand filtered and ozonated water) were analyzed using this protocol and the culture protocol associated with a PCR identification of typical colonies (table 2).

		Culture-PCR protocol	
		positive	negative
Direct detection	positive	5 (20%)	8 (32%)
using PCR	negative	0 (0%)	12 (48%)

Table 2 *Comparison between the direct PCR detection protocol and the culture-PCR protocol for the detection of Salmonella from water samples.*

The two methods reach the same result in 68% of cases. In 32% of cases, the direct detection results are positive, while the culture-PCR results are negative. The use of PCR as a direct detection method improves the detection by increasing its sensitivity and specificity. This conclusion is supported by the following observations. (i) no false positive result was observed during nested-PCR as proved by the correct response of the PCR-negative controls, and (ii) no positive result with the culture-PCR protocol was associated with a negative result with the direct PCR detection method. This new protocol reaches results within 24 hours instead of 4 days. Moreover, it is possible to detect several pathogenic bacteria from the same water sample with this method since it does not rely on the utilisation of selective media. Thus, we have optimised the PCR detection of two other pathogenic bacteria : at the species level (*Aeromonas hydrophila*) and at the sub-species level (enterohemorragic *Escherichia coli*)[4].

3.3 Viability of PCR detected bacteria

Unlike the culture techniques, PCR detection does not discriminate between viable and dead bacteria. We have developed an indirect approach in order to reach this goal when PCR is used as a direct detection method. A PCR assay is run for each water sample before and after a pre-enrichment step. An increase in the PCR signal after pre-enrichment proves the viability of the detected bacteria.

16 raw water samples were analyzed for the direct detection of the three targeted bacteria using this viability approach. The results are displayed in table 3.

	Salmonella	EHEC	*Aeromonas hydrophila*
Detection of viable bacteria	1	6	15
Detection of non viable bacteria	0	0	0
Negative result	15	10	1

Table 3 *Detection of viable pathogenic bacteria from raw water samples using PCR.*

When a positive result was obtained, the PCR signal was always more intensive after pre-enrichment. Therefore, the PCR detected bacteria were viable in all cases studied.

In this study, inhibitions of the PCR reaction were observed in some cases. Work is in progress in order to develop additional control tests in order to facilitate the interpretation of the results, in particular to compare inhibition levels of a sample treated before and after pre-enrichment.

4 CONCLUSIONS

This work shows that PCR is an excellent alternative to culture techniques. This technique allows an increase in both sensitivity and specificity of the detection and a decrease in the delay of analysis. When PCR is used as a direct detection method, several viable pathogenic bacteria can be detected from the same water sample. So far, PCR protocols have been developed for three pathogenic bacteria in our laboratory, we can envisage the PCR detection of other bacteria. Work is now in progress to develop control tests allowing to support the interpretation of the results.

5 REFERENCES

1. Mullis K.B., and Faloona F.A., Methods in Enzymol. 1987, **155**, 335-350.
2. Josephson K.L., Gerba C.P., and Pepper I.L., Appl. Environ. Microbiol., 1993, **59**, 3513-3515.
3. Way J.S., Josephson K.L., Pillai S.D., Abbaszadegan M., Gerba C.P., and Pepper I.L., Appl. Environ. Microbiol., 1993, **59**, 1473-1479.
4. Delabre K., Mennecart V., Joret J.C., and Cervantes P., 1997, Proceedings of the Water Quality Technology Conference, Denver, USA.

Rapid Methods in Water Chemistry

Mr Mark Smith

This section looks at some of the advances in rapid chemical measurements in waters.

The value of the water testing market for waters and effluents for rapid methods is estimated to be over £60m/annum in Europe alone. With the availability of a much wider range of methods than ten years ago and also improved performance, this is a rapidly growing area with an estimated annual growth rate of 8 – 10%.

The main advantages are:

- Instant at-site or at-line result is often possible.
- Significant cost-saving as often no or minimal laboratory facilities are required.
- Can often be used by relatively unskilled personnel 24 hours/day and 365 days/year.
- Usually less environmental impact than conventional laboratory testing.

It is very important that rapid methods are fully validated by the manufacturer prior to bringing the tests to market. This will include stability trials; batch to batch variability; robustness of the method; full performance testing involving a range of typical samples and spiked samples; documentation of potential interference effects; health and safety issues; full performance testing with a wide range of samples and finally safe disposal of any used test kits and associated reagents.

It is imperative that the method selected must be fit for the intended purpose. Also the sample presented to the analytical device/test kit must be representative of the substance being assessed. Care must be taken to ensure that the substance being determined it is not changed by the sampling process, sample container and during the storage period between the sampling process and the analysis.

The user will then need to ensure that the rapid method / test kit is suitable for his/her samples. The manufacturer should give advice on this and it is strongly advised that some comparison is also made with conventional laboratory analysis to ensure that the proposed method is fit for purpose and equivalent results are obtained prior to routine use of the test kits.

It is essential that all persons who will use the rapid methods / test kits are properly trained. They should then analyse a number of check samples to ensure that they can use the method in the intended manner and obtain fit for purpose results. They also need to be warned of any potential hazards and any other health and safety issues associated with the methods.

A system of documented quality control is considered essential. At the minimum this should consist of a blank and a suitable standard solution (preferably in a typical matrix) to be run on a regular basis with all results documented and preferably plotted on a Shewhart chart with warning and action limits. Participation in a relevant proficiency scheme (if available) is strongly recommended.

Recently a number of more 'exotic' rapid methods have been reported (e.g. AOX, total nitrogen; total petroleum hydrocarbons (TPH); various ecotoxicity tests etc.).

It will be interesting to see over the next ten years with the rapid advances in technology, what further rapid methods will be developed and also the percentage of work that will move out of the laboratory to the at-line or on-site location.

There would appear to be significant opportunities for proficiency testing schemes for rapid methods and test kits. It is hoped that these will be set up and that users will be encouraged to participate.

RAPID CHEMICAL ASSAYS BASED ON TEST KITS

Michael D. Buck

Hach Company
5600 Lindbergh Drive
Loveland CO 80538 USA

1 WHAT IS A TEST KIT

The ASTM defines a test kit (D5463-93 part 3.2.5) as a "commercially packaged collection of components that is intended to simplify the analytical function". Is a test kit to be portable? Is it to be used in the field away from power sources? There is a wide divergence of opinion even within our organization as to the proper definition of a test kit. We have, by consensus, arrived at a list of attributes of a test kit.

1. The test kit should be portable.
2. The test kit should have easy to use instructions and components.
3. The test kit should be self contained, having everything required to perform its analytical function.
4. The test kit must be usable in the field without external electricity.
5. The test kit must be accurate and reproducible, and provide a means to check the accuracy and operator technique.
6. The test kit should use stable, prepackaged, nonhazardous, safe reagents.
7. The analytical method should be fast.
8. The measurement system should be robust, not requiring customer calibration.
9. The test kit must be rugged and easily transportable by any means.
10. The precision, accuracy and interferences of the analytical method must be explained within the test kit instructions.
11. The test kit must be economical, fitting the application.

2 SELECTING A TEST KIT

There are a large number of manufacturers of test kits, and all publish specifications for their products. As a first step in the selection process, the user must decide where the kit is to be used, in the field or in a lab. Second, the user needs to define the sample type whether it be drinking water, wastewater etc. Third, what is the analyte and what is its likely concentration? Fourth, who will be using the test kit--a chemist or an untrained technician? Only after these four steps have been completed should one begin the fifth step, that of studying the various manufacturers' product specifications carefully.   Do not select a product that does not fit the application.   Thus, a manufacturer may offer as many as four different test kits for chlorine or hardness in water with these test kits differing in cost, accuracy and precision. If go/no-go testing is all that is required, why select a highly accurate and precise method when test papers might work as well?

     If the analyte concentrations are low, do not select a test kit with low resolution; i.e., if the sample contained 10 ppm hardness, one would not select a method where the minimum discernible step or division was 10 ppm.

     If the kit is to be used only in the field, do not select a kit requiring external electricity. If the potential user of the kit is not trained, do not allow the kit to be used without first requiring practice insuring that the user will use the kit properly. While the above would seem obvious, it is surprising how many customers order a test kit for the wrong reasons and are not satisfied.

3 PERFORMANCE OF EXISTING TEST KITS

Downes *et al.*[1] evaluated the performance of a cyanuric acid test kit, which used a visual turbidimetric method, versus an instrumental turbidimetric method and a UV method. They concluded that although the test kit was not as precise as the instrumental methods, it was nevertheless sufficiently accurate for poolside monitoring.

     Spokes and Bradley[2] evaluated the performance of the CHEMETS lead test kit based on dithizone versus atomic absorbtion spectrophotometry. Eighty-four samples were compared, and the authors obtained an R^2 of 0.951 between the two methods. Unfortunately, interferences were not evaluated.

     Schock[3] evaluated the performance of a lead test kit based on a porphyrin method versus graphite furnace atomic absorbtion spectrophotometry. Schock concluded that at concentrations between 10 and 100 µg/L lead, the absolute precision of the test kit was equal to ±3 µg/L. Schock further concluded that both methods had an equivalent precision at lead levels above 40 µg/L.

     Ormaza-Gonzales *et al.*[4] evaluated the performance of a nitrate, a nitrite and a phosphate test kit versus EPA standard methods. The authors concluded that the results were comparable at levels of nitrite and phosphate above 1 µM for both fresh and sea water. Nitrate results were comparable in fresh water but required standard additions in sea water, as the kit instructions noted.

     Beenakkers *et al.*[5] studied the recovery of 2,4-D of two enzyme-linked-immunoassay (ELISA) test kits versus GC-MS. The authors reported that one of the test kits offered excellent recovery and a low detection limit in river water, less than 0.05 µg/L. The second kit did not perform as well, with a detection limit greater than 0.1 µg/L. The analysis time was less than 2 hours per assay. Coefficient of variation was between 2 and 3%.

     Dankwardt *et al.*[6] studied the recovery of atrazine of a magnetic particle-based ELISA test kit in natural waters and soils versus GC and a microtitre plate method. The authors reported detection limits of 0.04 ppb for the test kit and 0.02 ppb for the GC method. Coefficients of variations of less than 10% were observed. Response toward prometryne, desethylatrazine and ametryne were 25, 15 and 12% that of atrazine. Analysis time was reported at less than one hour.

4 TEST KITS OF THE FUTURE

Test kits of the future will be smaller, faster, more sensitive, and more precise than test kits of the past. Colorimetric and titrametric methods will still be with us but will be used primarily in less expensive, slower kits. Test kits of the future will incorporate newer instrumental methods which have been made possible by the advent of micro-electronics and the very creative work of chemists like John Hart, Stephen Wring, Joseph Wang, to mention but a few. Newer test kits will see the use of small, single-use sensors and hand-held instruments to perform potentiometric, amperometric and optical methods of analysis. In the last 15 years, more than 500 papers have been written about chemically-modified sensors. For example, the single most common method of field analysis performed is that for blood glucose. In one of the common test kits, illustrated below (Figure 1), sold for blood glucose, a chemically modified amperometric sensor is used. Examples of some of these methods are described below.

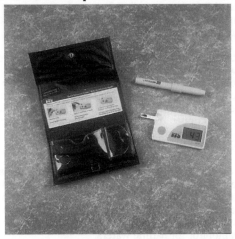

Figure 1 *Blood Glucose Analysis System. (Note the drop of blood on sensor strip.)*

5 CHEMICALLY MODIFIED SENSORS, AMPEROMETRIC

Ag   Li *et al.*[7] were able to determine silver at levels of $5-10^{-10}$ to 1×10^{-7} M by accumulating silver at the surface of a carbon paste electrode for 5 minutes which had been modified with a poly thioether chelating resin. After medium exchange, the accumulated silver was reduced to metal at -0.4 V (Volt), after which the potential scanned from -0.4 V to +0.55 V. Silver was reoxidized with a peak current at +0.35 V. The method was applied to the determination of silver in wastewater.

Au   Gao *et al.*[8] described the determination of gold at levels of 1×10^{-8} to 1×10^{-6} M with a 5-minute accumulation time. The sensor was a carbon paste electrode which had been modified with an ammino-isopropylmercapto chelating resin. After medium exchange, the accumulated gold was reduced to metal at -0.40 V. The potential was scanned from -0.40 to +0.80 V with gold being reoxidized at +0.58 V. The method was applied to the determination of gold in minerals, copper and anode mud.

Cd   Zhang *et al.*[9] described the determination of cadmium at levels of 1×10^{-8} to 1×10^{-5} M with an accumulation time of 5 minutes. The sensor was a carbon paste electrode which had been modified with 8-hydroxyquinoline. Cadmium was accumulated at -0.6 V for 60 seconds and reduced to metal by scanning from -0.6 V to -1.0 V. The maximum reduction current occurred at -0.78 V. The method was applied to natural waters.

Co   Gao *et al.*[10] determined cobalt levels of 1×10^{-7} to 9×10^{-6} M with a detection limit of 8×10^{-8} M. The sensor was a carbon paste electrode modified with nafion and 1,10-phenanthroline. Cobalt was accumulated for 120 seconds open circuit, then reduced to metal at -0.3 V. The potential was scanned from -0.3 V and +0.7 V with cobalt being oxidized with a peak current at +0.12 V.

Fe   Compagnone *et al.*[11] described the determination of Fe^{+3} at levels of
          1×10^{-7} to 1×10^{-6} M. The sensor was a carbon paste electrode modified
          with 2-methyl-3-hydroxypyridine-2-one. Iron was accumulated for 120
          seconds open circuit as Fe^{+2}. The potential was scanned from 0 to
          +0.6 V with an oxidation peak at +0.41 V.

Cu   Peng *et al.*[12] described the determination of copper at levels of 1×10^{-8} to
          1×10^{-6} N with a detection limit of 3×10^{-10} M. The sensor was a carbon
          paste electrode modified with α-benzoinoxime. Copper was collected at
          the surface of the sensor for 120 seconds open circuited. The
          accumulated copper was reduced to copper metal at -0.8 V for 60
          seconds. The copper metal was then reoxidized by scanning between
          -0.8 to +0.4 V with a maximum current peak at +0.1 V. The method was
          applied to industrial wastewater and anode mud.

Hg   Wang *et al.*[13] described a remote sensor for the determination of mercury
          with a detection limit of 2×10^{-9} M with a 5-minute accumulation time. The
          sensor was a gold wire with surface covered with agarose. Mercury was
          accumulated at +0.2 V and potentiometrically stripped at a constant
          0.5 μA. Mercury was stripped at +0.4 V. The sensor was designed as a
          continuous monitor for natural waters.

Ni   Wang *et al.*[14], extending the work of Baldwin *et al.*[15] on carbon paste
          electrodes, described the determination of nickel at levels of 8×10^{-8} to
          2×10^{-6} M using a disposable, screen-painted carbon ink sensor. The
          carbon ink was modified with dimethylgloxime. Nickel was accumulated
          at an open circuit on the surface of the electrode for 60 seconds and then
          reduced to metal by scanning the potential from -0.8 V to -1.4. The
          concentrations of nickel was related to the reduction current peak at
          -1.1 V. The method was applied to river water.

Tl   Cai *et al.*[16] described the determination of thallium at levels of 5×10^{-10} to
          1.6×10^{-5} M using a carbon paste electrode modified with 8-hydroxy-
          quinoline. Thallium was accumulated for 120 seconds at the surface of
          the electrode, open circuit. The accumulated Tl^{+3} was reduced to metal at
          -1.2 V and the metal reoxidized by scanning the potential from -1.2 V to
          -0.7 V. Thallium concentration was related to the current at a sharp peak
          at -0.89 V. The method was applied to sea water and human urine.

    In general, methods that use chemically modified sensors gain their selectivity by
the choice of the ligand used to modify the electrode surface, collecting the metal in
the open-circuit mode with the use of masking agents. Carbon paste electrodes are a
research tool and are not likely to be used in day-to-day analysis. However, sensors
formed by screen printing similar to Wang[14] will be used in routine analysis similar to
the glucose sensor shown above.

6 IMMUNO SENSORS, AMPEROMETRIC AND POTENTIOMETRIC

6.1 Polychlorinated Biphenyls (PCBs)

    Bender *et al.*[17] described a direct electrochemical immuno sensor for the
determination of PCBs in industrial effluents and seafood plant effluents. The assay is
based on the measurement of the current due to the specific binding between PCB and
anti-PCB antibody-immobilized conducting graphite ink. The sensor gave linear
response from 0.3 to 100 μg/mL.

6.2 *Cryptosporidium Parvum*

Wang *et al.*[18] demonstrated a screen printed carbon sensor for the determination of DNA sequences from the waterborne pathogen *cryptosporidium*. The sensor relies on the immobilization of a 38-mer oligonucleotide unique to *cryptosporidium* DNA onto the carbon sensor and employs a highly sensitive chronopotentiometric mode for monitoring the hybridization event. The method was applied to untreated drinking water and river water. Determinations required less than 25 minutes.

In the same manner Wang[19] developed a screen printed carbon sensor for the determination of the DNA sequence of *E. coli* and applied it to drinking water and river water samples.

6.3 *E. coli* O157:H7

Gehring *et al.*[20] described a light-addressable potentiometric sensor (LAPS) for the determination of *E. coli* O157:H7. The samples were treated with 20 ng of biotrinylated Goat-anti *E. coli*)157:H7, fluorescein-labeled Goat-anti *E. coli* O157:H7 and urease-labeled anti-fluorescein antibody conjugate. The mixture was mixed and incubated for 30 minutes at 25°C. Two µg streptavidin was added, mixed and filtered through a Molecular Devices "threshold stick". The stick was placed in a reader chamber where light illuminated the bacteria captured on the filter stick and caused the pH to change at the surface of the stick. The rate of change of pH was proportional to the number of bacteria captured on the stick. Typical analysis time was 40 minutes.

7 ENZYME MODIFIED AMPEROMETRIC SENSORS

John Hart *et al.*[21] developed a method for the measurement of organophosphate pesticides using a screen printed disposable carbon sensor modified with acetylcholinesterase and cobalt phthalacyanine. The sensor was incubated with a buffer with an applied potential of +0.1 V. After 9 minutes, acetylthiocholine was added and a steady state current obtained nine minutes after that. The sample was then added and the rate of current decrease was related to the inhibition of the enzyme by the pesticide and thus to its concentration.

8 IMMOBILIZED BACTERIA SENSORS

8.1 BOD

Tanaka *et al.*[22] described a sensor for BOD which has since become a Japanese standard, (JIS) K3602. The sensor is a layer of *trichosporon cutaneum* L-1 strain immobilized in polyvinyl alcohol sandwiched between 2 membrane filters. The sandwiched layer was fixed in place over a Clark dissolved oxygen electrode. When the sensor is placed in a sample containing dissolved organic material, a portion of that organic material will diffuse into the immobilized bacteria layer and be consumed by the bacteria. The consumption of the organic materials requires oxygen by the bacteria. Thus the oxygen measured by the Clark electrode will decrease. The decrease in oxygen is related to the BOD_5 of the sample. Typical analysis time is 40 minutes per sample.

9 OPTICAL SENSORS

9.1 BOD

Reynolds *et al.*[23] developed a fluorescence procedure to provide an instant surrogate indicator of BOD. He used 280 nm excitation and 340 nm emission. To compensate for attenuation of the incident light by the sample, the fluorescence signal was normalized by the water Raman signal. The normalized fluorescence signal fit a straight line with correlation coefficients of 0.97 to 0.89, depending upon the sample source. Since COD and TOC are surrogate methods for BOD and require one or more hours analysis time, the use of site-specific fluorescence as a process control method should be more than welcome.

9.2 Tetracycline

Lin *et al.*[24] developed a fluorescence sensor for the determination of tetracycline. The sensor membrane was an anthracene containing copolymer, methyl- and butylmethacrylate and dioctyl sebacate. Anthracene, within the polymer membrane, excited at 369.5 nm, fluoresces at 412.8 nm. When the anthracene is exposed to tetracycline the fluorescence is decreased, and the amount of decrease is related to the concentration of tetracycline in the sample. The sensor was claimed to have a long lifetime, 24 hours, under continuous exposure and was applied to pharmaceuticals and urine.

9.3 Ammonium Ion

Simon *et al.*[25] developed a flow-through sensor for the determination of NH_4^+. The sensor was based on a plasticized PVC membrane containing nonactin and ETH 5294, a H^+-selective neutral chromoionophore. Nonactin in the membrane would equilibrate with NH_4^+ in the sample and cause the H+-selective chromoionphore to change its color from blue to orange. The change in color is proportional to the NH_4^+ concentration. The time constant of the sensor was 30 seconds.

9.4 Cd, Pb and Hg

Czolk *et al.*[26] described a reflectance sensor based on complexation of Cd, Pb and Hg by 5, 10, 15, 20-Tetra (p-sulfonatophenyl) porphyrin covalently immobilized in a polymer matrix. The porphyrin used formed colored chelates absorbing at different wavelengths for the three metals. By deconvoluting the reflectance spectra the authors were able to determine the concentrations of each of the three metals.

10 CONCLUSION

The above technology should serve to illustrate the direction of analytical chemistry and therefore the direction of test kits. The use of chemically modified sensors, immuno-sensors, enzyme modified sensors, and optical sensors will open the range of analysis available for use in test kits removing the distinction between laboratory and field analysis. Instruments are becoming smaller, demanding less power to operate, and thus becoming more portable. A complete spectrophotometer, the size of a pack of cigarettes, is already available. Anodic stripping voltametric portable test kits, as well as potentiometric ion-selective electrode portable test kits, are available. Analytical chemistry is moving toward the increased use of sensors, and hand-held instruments and test kits will move in the same direction.

References

1. C. J. Downes, *et al.*, *Water Res.*, 1984, **18**, 277.
2. G. N. Spokes, *et al.*, ASTM STP 1102, 1991.
3. M. R. Schock, *J. Am. Water Works Assoc.*, 1993, **85**, 90.
4. Ormaza-Gonzales and Villaba-Flor, *Water Res.*, 1994, **28**, 2223.
5. Beenakkers, *et al.*, *Water*, 1995, **28**, 624.
6. A. Dankwardt, *et al.*, *Acta Hydrochim.Hydrobiol.*, 1993, **21**, 110.
7. P. Li, *Anal. Chim. Acta*, 1990, **229**, 213.
8. Z. Gao, *Anal. Chim. Acta*, 1990, **232**, 367.
9. Z. Zhang, *Lihua Jianyan, Huaxe Fence*, 1997, **33**, 506.
10. Z. Gao, Fres. J. Anal. Chem., 1991, **339**, 137.
11. D. Compagnone, *et al.*, *Sensors and Actuators B*, 1992, 7, 549.
12. J. Peng, *et al.*, *Mikrochim. Acta*, 1996, **122**, 125.
13. J. Wang, *Electroanalysis*, 1998, **10**, 399.
14. J. Wang, *Electroanalysis*, 1996, **8**, 635.
15. R. Baldwin, *Anal. Chem.*, 1996, **58**, 1790.
16. Q. Cai, *Analyst*, 1995, **120**, 1047.
17. S. Bender, *et al.*, *Environ, Sci. and Tech.*, 1998, **32**, 788.
18. J. Wang, *et al.*, *Talanta*, 1997, **44**, 2003.
19. J. Wang, *et al.*, *Electroanalysis*, 1997, **9**, 395.
20. A. G. Gehring, LAPS Sensor, *Anal. Biochem.*, 1998, **258**, 293.
21. J. P. Hart, *Anal. Proc. Incl. Anal. Comm.*, 1994, **31**, 333.
22. I. Karube, *Pure Appl. Chem.*, 1987, **59**, 545.
23. D. M. Reynolds, *et al.*, *Water Res.*, 1997, **31**, 2012.
24. W. Liu, *et al.*, *Analyst*, 1998, **123**, 365.
25. W. Simon, *et al.*, *Anal. Sci.*, 1989, **5**, 557.
26. R. Czolk, *et al.*, *Sensors and Actuators B*, 1992, 7, 540.

A CONTRIBUTION TO CONVENIENCE FOR ENZYME-BASED ASSAYS OF PESTICIDES IN WATER

A. L. Hart and W. A. Collier

AgResearch Grasslands
Private Bag 11008
Palmerston North
New Zealand

1 INTRODUCTION

Rapid assays are required for organo-phosphate and carbamate pesticides. Many assays for these compounds use cholinesterases,[1] but can be quite time consuming. The inhibition of cholinesterase by pesticides is a kinetic process.[2] At low pesticide concentrations (e.g. 10^{-8} M) long incubation times (e.g. 30-45 minutes) can be required to detect the inhibition of enzyme activity.[1,3]

One way to circumvent the time required to perform assays sequentially would be to arrange them in parallel. Further, if the period between loading the arrays and measuring enzyme activity could be extended past conventional incubation times, without disturbing dose-response relationships, then this extension may prove convenient not only in laboratory tests but also in field testing.

The idea that there could be a long interval between exposure and testing for assays based on cholinesterases, and that this interval could be a substitute for the conventional incubation process was put forward in a recent paper.[4] Further evidence is provided here for the validity of this approach.

2 METHODS

The assays were based on electrochemical measurement of cholinesterase activity. The components of an anode of a three-electrode electrochemical cell were assembled by screen-printing. The anode contained carbon and cobalt phthalocyanine. Cholinesterase embedded in a matrix of lactitol and a quartenized vinyl pyrrolidone/ dimethylaminoethyl methacrylate copolymer (GafQuat 755N) was screen-printed onto the anode. A thin outer layer of polyurethane was applied with an air-jet.[3,4]

Droplets of pesticide solution or water (3 or 9 μl) were applied to the electrodes and allowed to dry (about an hour). The time elapsing between pesticide application and commencement of testing was about 17 hours.

Cholinesterase activity was measured by recording the current, generated in the presence of 0.25 mM acetyl thiocholine, a potential of 100 mV vs Ag/AgCl (a platinum wire formed the auxiliary electrode).

The pesticides used were two organo-phosphates, malathion and acephate, a carbamate, methomyl, and commercial formulations containing these: Maldison, Orthene and Lannate. The latter were diluted to give multiples of the recommended dilution (RD) for application, and the pure compounds were diluted to give molar concentrations equal to the concentration of the active ingredient in the various dilutions of the commercial products. Maldison at 1 RD contains 6.1 mM malathion; Orthene, 6.1 mM acephate; and Lannate, 2.2 mM methomyl.

Droplets (3 μl) of Maldison and Orthene at 10^{-5} to 10 times RD, and of malathion and acephate at the corresponding molar concentrations were applied to electrodes. Droplets (3 μl) of Lannate and methomyl were applied in a similar experiment. In a further experiment, two levels of enzyme were used in the construction of the electrodes, 0.25 and 1.0 mg per 5 ml of enzyme matrix. Maldison and Lannate were applied at 10^{-5} to 1 RD using 3 and 9 μl droplets.

3 RESULTS

When exposed to single doses of pure and commercial formulations of malathion and acephate, there was a generally smooth decline in current (Figure 1). Currents from electrodes exposed to pesticides were not significantly different from those exposed to water until the concentration reached 6.1 x 10^{-5} M malathion, 10^{-1} RD Maldison and Orthene, and 4.1 x 10^{-3} M acephate.

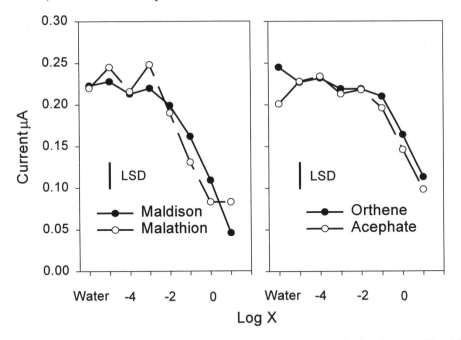

Figure 1 *Response of electrodes to organo-phosphates. X is a multiple of RD or 6.1 mM (malathion) or 4.1 mM (acephate)*

The currents from electrodes exposed to methomyl also bore a regular relationship to concentration, the response becoming significantly different from that to water at 2.2 x 10^{-7} M methomyl and 10^{-4} RD Lannate. (Figure 2).

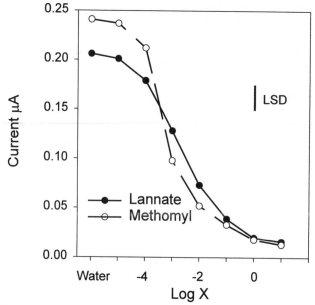

Figure 2 *Response of electrodes to a carbamate. X is a multiple of RD or 2.2 mM (methomyl)*

The mean of the currents generated by 'high enzyme' electrodes was about twice that from 'low enzyme' electrodes (0.21 vs. 0.10 μA; Figure 3). For both high and low enzyme electrodes, the currents generated declined with increasing pesticide concentration the decline being more severe for Lannate than Maldison. An effect of droplet size was only apparent at the highest concentrations where currents from electrodes exposed to 9 μl droplets were lower than from those exposed to 3 μl droplets.

4 DISCUSSION

In that the responses to pesticide concentration were regular, the results bear out the previous claim that for cholinesterase contained in a suitable matrix, exposure and testing can be separated by quite substantial periods of time. These periods are long enough to allow for the required incubation between pesticide and inhibitor,[2] and for any logistical arrangements. Tests carried out in this way could also be arranged as a parallel array.

The mixture of a cationic polyelectrolyte and a sugar alcohol which formed the enzyme matrix has been noted for its ability to confer stability on enzymes in (bio)chemical sensors.[5-7] The enhanced stability is thought to arise from increased hydrogen and electrostatic bonding.[8] This network of bonds may prevent adverse effects arising from local changes in water potential.

Other sensors based on cholinesterase are capable of detecting pesticides in the range 10^{-7} to 10^{-9} M.[9,10] The electrodes described here were not quite as sensitive, the maximum sensitivity being of the order of 10^{-7} M with methomyl. At the greatest dilutions of the pesticides there may have been insufficient penetration of the inhibitor into the enzyme layer before evaporation of the solvent.

The small effect of increasing the droplet size to 9 µl may have arisen from the fact that it represents only a small change in dose in the context of the large range of concentrations.

The responses of the electrodes to pesticide concentration were not always completely smooth. These irregularities may have arisen from variability in the structure of the electrodes induced during manufacture. Variation in the thickness of the enzyme and other layers can have a large effect on the performance of enzyme-based sensors. [11]

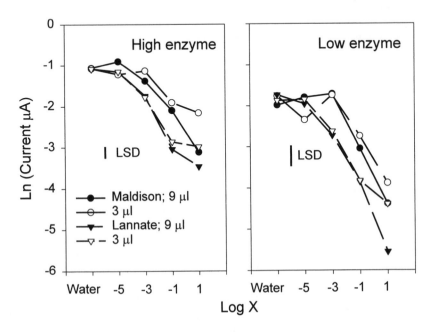

Figure 3 *Response of electrodes to variation in enzyme level and pesticide dose. X is a multiple of RD*

References

1. P. Skladal and M. Mascini, *Biosens. & Bioelect.*, 1992, **7**, 335.
2. W. N. Aldridge, *Biochem. J.*, 1950, **46**, 451.
3. A. L. Hart, W. A. Collier and D. Janssen. *Biosens. & Bioelect.*, 1997, **12**, 645.
4. A. L. Hart and W. A. Collier, *Sens. & Act.* B, 1998, **53**, 111.
5. W. A. Collier and A. L. Hart, *Australasian Biotech.*, 1997, **7**, 93.
6. T. D. Gibson, J. N. Hulbert and J. R.Woodward, *Anal. Chim. Acta*, 1993, **279**, 185.
7. J. J. Rippeth, T. D. Gibson, J. P. Hart, I. C. Hartley, and G. Nelson, G. *Analyst,* 1997, **122**, 1425.
8. T. D. Gibson and J. R. Woodward, 'Biosensors and Chemical Sensors', American Chemical Society, Washington, 1992, Vol. 487, p.40.
9. I. C. Hartley and J. P. Hart, *Analyt. Proc. Including Analyt. Commun.* 1994, **31**, 333.
10. P. Skladal, M. Fiala and J. Krejci, *Intern. J. Environ. Anal. Chem.*, 1996, **65**, 139.
11. E. A. H. Hall, J. J. Gooding, C. E. Hall, *Mikrochim. Acta*, 1995, **121**, 119.

WATER QUALITY CONTROL IN THE NETHERLANDS: SELECTION AND EVALUATION OF INDICATIVE METHODS

E.P. Meulenberg

ELTI Support, Drieskensacker 12-10, 6546 MH Nijmegen, The Netherlands

1 INTRODUCTION

Due to national and international norms and regulations, an extensive water quality control program exists in the Netherlands. Both surface water and drinking water has to be monitored for numerous parameters and contaminants, including pesticides. The EC norm for pesticides in drinking water and surface water used for the production thereof is 0.1 µg/L for single compounds and 0.5 µg/L for the sum of pesticides. Compliance of these norms for a prolonged period leads to a stop of surface water intake and consequently to a shortage of drinking water for a great part of the country.

The conventional methods used for the analysis of pesticides are GC, HPLC and MS, which are laborious, time-consuming and expensive. Taking into account that at least a great proportion of the samples is negative for particular pesticides, especially those that are applied only in certain periods of the year, such as herbicides, there exists a strong need for alternative screening methods. In an extensive inventarisation several candidate screening methods were evaluated. Deskresearch revealed that toxicity testing, HPLC fingerprinting and immunoassay may present such indicative methods. Further studies involved the assessment of the applicability of these methods in water quality control based on various criteria. The most important property desired was the ability to discriminate between contaminated and uncontaminated samples. The overall result was that the immunoassay was regarded as superior for reasons of rapidity, selectivity, sensitivity and cost-effectiveness.

In order to introduce the immunoassay as an analytical tool for screening purposes in water quality control, several commercially available ELISA kits for pesticides were further evaluated on laboratory scale and validated using real surface water samples. ELISA results were compared with reference values, if appropriate. Especially polar pesticides such as triazines, urea herbidies and carbendazim appear to be suitable target compounds for screening by ELISA.

2 METHODS AND MATERIALS

2.1 Methods

2.1.1 Deskresearch. An inventarisation of available assays useful as indicative methods in water quality control was made by searching the literature, including on-line data bases, and on the basis of in-house experience. Assays were evaluated for general availability, applicability, sensitivity, robustness, and cost-effectiveness. The most promising assays were included in further investigation for evaluation and validation.

2.1.2 HPLC-fingerprint. On-line SPE-HPLC was performed by applying a 15 mL sample onto a non-specific PLRP-s column. The eluate was passed to an analytical column and UV detection was conducted at 215 nm. From the chromatogram the peak sum (PS) and total determined organic matter (TDOM) were calculated.

2.1.3 ELISA. Each assay was performed according to the instructions of the manufacturer. Conventional parameters were determined, including sensitivity, linearity, precision, accuracy, cross-reactivity, and matrix effects.

2.1.4 Reference methods. Samples, unknown or spiked with several combinations of pesticides were also measured by reference analysis. Urea-herbicides were analysed by HPLC/DAD; triazines with GC-MS, carbendazim by SPE-HPLC/DAD; and cyclodienes by GC/ECD.

2.1.5 Calculation and Statistics Standard curves and unknown samples in ELISA were calculated by using 4-parametric logit-log transformation. Statistical analysis of the data was performed by linear regression and non-parametric Wilcoxon Signed Rank Test.

2.2 Materials

ELISA microtiter plate kits for isoproturon, urea herbicides, carbendazim, cyclodienes, and tube kits for triazines were purchased from SDI Europe (Hampshire, UK). Surface water was sampled at various locations into glass bottles and at 4°C. Before use, part of the samples were pretreated by filtration (0.45 μm filter, Millipore) or by sedimentation to remove particulate matter. Samples for evaluation studies were made by addition of pesticides at several concentrations and in different combinations to tap water or surface water. Standard pesticides were of high purity as used in accredited chemical analysis; all other chemicals and reagents were analytical grade.

3 RESULTS

3.1 Deskresearch

Literature search combined with in-house experience resulted in the selection of several candidate assays that may be useful as indicative methods for water quality control. Among the toxicity tests, four commercially available tests did meet the initial criteria: ISO 6341 Acute Dapnia test; IQ-test Daphnia magna; Thamnotox F-kit; and Thamnotox Fluo-kit[1]. An assay based on conventional HPLC was the HPLC-fingerprint method[2], wherein PS and or the ratio PS/TDOM was used as a sum parameter. It was shown that using a 15 mL sample a run for 40 minutes and detection of peaks at 215 nm yielded a sum parameter that could be used to discriminate contaminated from uncomtaminated samples. This method was especially useful for polair pesticides. With regard to immunoassay, among the various kits commercially available, initially four pesticide kits were selected as suitable for further research on the basis of occurence of target compounds and specifications included in the kit insert. Pesticides selected comprised triazines, isoproturon, urea-herbicides and cyclodienes.

3.2   Evaluation

The methods found by desk reseach were further evaluated on the basis of criteria such as applicability in the sense of simplicity, rapidity and costs, and of indicative potential, i.e. useful in routine screening programs. Toxicity tests were considered as too laborious due to the pretreatment of a very large sample volume in order to achieve a suitable detection limit. Furthermore, the reliability of the results may be insufficient for introduction at unexperienced laboratories. HPLC-fingerprint seemed suitable to obtain an overall indication of contamination, whereas immunoassay may be used either for single-compound or group selective screening purposes. Both these latter methods were included in further evaluation studies.

The performance of the methods was evaluated by first determining general parameters such as detection limit, precision, linearity, recovery, matrix effects. Additionally, the accuracy of the respective methods was established by measuring spiked surface water samples (n=15) and comparing the values found with those of the reference analysis. The results for the HPLC-fingerprint are given in Table 1; ELISA results are presented in Table 2.

To assess the reliability of both methods with regard to the potential to discriminate contaminated form uncontaminated samples, field samples unspiked or spiked with several combinations of pesticides were first analysed by conventional analytical methods. Then they

Table 1. *Performance parameters of HPLC-fingerprint*

Parameter	Determined	Value/Score
Reproducibility	PS TDOM	5.3 % 3.6 %
Background	PS; TDOM	< 5 %
Memory effect	Relative increase in background	nihil
Matrix effects	PS or PS/TDOM expected to found	occasionally
Reproducibilty RT: standards(12)/spikes(6)	C.V.	< 0.25 % (< 0.05 min)
Recovery: 0.2-0.5 µg/L 1.0-2.0 µg/L	C.V.	 > 80 % > 75 %
Quantitive reproducibility: 0.2-0.5 µg/L 1.0-2.0 µg/L	C.V.	 < 5 % < 5 %
Linearity (n=7)	Correlation coefficient	> 0.94

Sensitivity:
  surface water
  ultra pure water
	Detection limit	
		0.02 - 0.08 µg/L
		0.02 - 0.06 µg/L

were subjected to HPLC-fingerprinting and immunoassay. In Table 3 the results are given for the samples containing polar pesticides. Because cyclodienes required a alternative analytical method and were not subjected to HPLC-fingerprint, comparison of the ELISA results versus GC are given in Table 4.

Table 2. *Performance parameters of immunoassays*

ELISA kit	Parameter	Determined	Values/score
Triazines Isoproturon Urea-herbicides Cyclodienes	Sensitivity	Detection limit	0.03 µg/L 0.01 µg/L 0.006 µg/L 0.2 µg/L
Triazines Isoproturon Urea-herbicides Cyclodienes	Precision	Intra-assay CV	3.5 - 5.0 % (n=5) 3.0 - 9.0 % (n=5) 10.5 % (n=5) 3.2 - 16 % (n=5)
Triazines Isoproturon Urea-herbicides Cyclodienes	Precision	Inter-assay CV	8 - 11 % (n=6) 9 - 12 % (n=6) 10 - 17 % (n=5) 51 - 58 % (n=2)
Triazines Isoproturon Urea-herbicides Cyclodienes	Linearity	Correlation coeff.	0.996-0.999 (n=3) 0.992-0.989 (n=2) 0.991-0.982 (n=2) 0.987 1.000 (n=3)
Triazines Isoproturon Urea-herbicides Cyclodienes	Matrix effects	Concentrations expected vs. found	F: -; S: ± F: ±; S: ± F: +; S: ± F: ++; S: ++

Legend: F = filtration; S = sedimentation; -, ±, +, ++ = scores for the increasing effect of sample pretreatment.

Table 3. *Comparison of conventional analysis, HPLC-fingerprint and ELISA*

Sample ID	Compounds	Conc (a) (μg/L)	Conc (b) (μg/L)	Conc. (c) (μg/L)	HPLC-fingerprint PS-TDOM-ratio
PP03A	---	---	---	<0.01-0.15-0.03	340 - 22399 - 0.015
PP01A	Diuron	0.15	0.17	0.39	441 - 20861 - 0.021
PP04A	Diuron	0.15	0.27	0.53	510 - 22838 - 0.022
	Isoproturon	0.15	0.15	0.15	
PP09B	---	---	---	<0.01-0.16-0.04	613 - 17824 - 0.034
PP06B	Atrazine	0.17	0.31	0.43	1336 - 17620 - 0.076
	Simazine	1.17			
PP07B	Diuron	0.15	0.27	0.4	1427 - 17862 - 0.080
	Carbendazim	0.61			
PP05C	---	---	---	<0.01-0.01-<0.03	175 - 4047 - 0.043
PP08C	Diuron	0.15	0.27	0.38	301 - 4057 - 0.074
	Vinclozolin	0.68			
PP13C	Atrazine	0.17	0.31	0.44	598 - 3903 - 0.153
	Simazine	1.17			
	Mevinphos	0.28			
	Vinclozolin	0.68			
	Metribuzine	0.35			
	Pirimicarb	0.20			
	Metolachlor	1.20			
PP14D	---	---	---	<0.01-0.16-0.04	438 - 6142 - 0.073
PP11D	Diuron	0.15	0.35	0.42	471 - 6360 - 0.074
	Mevinphos	0.28			
PP12E	---	---	---	<0.01-0.07-0.11	385 - 180×10^3 - 0.0021
PP02E	Isoproturon	0.15	0.15	0.15	592 - 166×10^3 - 0.0036
PP10E	Diuron	0.15	0.59	0.49	588 - 176×10^3 - 0.0033
	Chloridazon	0.17			
PP15E	Atrazine	0.17	0.31	0.59	626 - 180×10^3 - 0.0035
	Simazine	1.17			
	Mevinphos	0.28			
	Vinclozolin	0.68			
	Metribuzine	0.35			
	Pirimicarb	0.20			
	Metolachlor	1.20			

Legend: Samples were collected at five different sites A, B, C, D, E; Compounds = pesticides added; Conc. (a) = concentration added; Conc. (b) = concentration expected based on cross-reactivity in the respective kits; Conc. (c) = concentration found, values given for unspiked samples are those found in the respective kits; PS = peak sum; TDOM = total determined organic matter; ratio = PS/TDOM.

Table 4. *Comparison of results of GC/ECD and ELISA*

Sample ID	Compounds	Conc. (a) (µg/L)	Conc. (b) (µg/L)	Conc. (c) (µg/L)
C03US	---	---	---	2.4
C01US	Chlordane	15.1	15.1	36
C05US	Endrin	0.5	5.3	13.2
C09US	Chlordane	15.1	41	79.2
	Endrin	0.5		
	alfa-Endosulfan	2.3		
	Heptachlor	9.8		
C06RA	---	---	---	3.2
C04RA	Dieldrin	5.9	6.6	17.2
C07RA	alfa-Endosulfan	2.3	11.7	23.7
C11RA	Chlordane	15.1	41	63.8
	Endrin	0.5		
	alfa-Endosulfan	2.3		
	Heptachlor	9.8		
C10RB	---	---	---	4
C12RB	Chlordane	15.1	41	41.2
	Endrin	0.5		
	alfa-Endosulfan	2.3		
	Heptachlor	9.8		
C13RD	---	---	---	4.8
C02RD	Aldrin	33	11.9	37.2
C08RD	Heptachlor	9.8	8.9	43.7
	Chlordane	15.1	41	64.3
C14RD	Endrin	0.5		
	alfa-Endosulfan	2.3		
	Heptachlor	9.8		
C15WO	---	---	---	3.2

Legend: see Table 3; samples were collected at five different sites US, RA, RB, RD, WO; --- indicates unspiked samples

3.3 Validation

On the basis of the results obtained in the evaluation study, the immunoassay was selected for further validation and implementation in laboratories for water quality control. From the target compounds the cyclodienes were replaced by carbendazim, because cyclodienes are rather apolar with a tendency to adhere to particulate matter, and immunoassay results are strongly influenced by matrix effects. On the other hand, available data on rate of application and occurence in surface water of carbendazim and the availability of an immunoassay kit suggested this to be an useful alternative target compound.

For the validaton study surface water was sampled (n=10) at various sites, divided over three portions and analysed for triazines, urea-herbicides, and carbendazim both with immunoassays and conventional methods. The immunoassays were performed at three

Rapid Detection Assays for Food and Water

different unexperienced laboratories and the concentrations found were compared to those determined by GC/NPD and HPLC, respectively. The results are given in Table 5.

Table 5. *Validation of immunoassays versus conventional analytical methods.*

IA Kit	Atrazine		Urea-herbicides		Isoproturon		Carbendazim	
Sample ID	IA (n=6)	GC/MS	IA (n=6)	HPLC	IA (n=5)	HPLC	IA (n=6)	HPLC
M01	0.36±0.05	0.55	0.01±0.005	0.09	0.005±0.004	< 0.05	0.04±0.02	0.1
M02	> 1.0	1.8	> 2.0	3.24	0.056±0.011	< 0.05	0.13±0.01	< 0.05
M06	0.64±0.12	1.09	0.84±0.11	0.45	0.015±0.007	< 0.05	0.70±0.17	0.72
M11	0.23±0.20	0.38	0.96±0.15	0.58	0.009±0.012	< 0.05	0.50±0.14	0.56
M12	0.23±0.03	0.33	0.70±0.14	0.62	0.016±0.013	< 0.05	1.12±0.24	1.01
M13	0.26±0.26	1.02	0.25±0.07	0.25	0.010±0.009	< 0.05	1.26±0.23	1.02
M15	0.33±0.40	0.93	0.94±0.24	0.55	0.008±0.008	< 0.05	0.82±0.23	0.7
M16	0.28±0.31	1.8	1.78±0.13	1.4	0.007±0.008	< 0.05	0.54±0.13	0.94
M17	0.27±0.32	0.65	0.61±0.12	0.31	0.042±0.030	< 0.05	0.79±0.17	0.67
M19	0.27±0.39	0.21	0.16±0.06	0.12	0.007±0.008	< 0.05	0.51±0.18	0.34
D.L.	0.01-0.3	0.01-0.03	0.01-0.04	0.03-0.2	0.001-0.02	0.05	0.04-0.11	0.05-0.2

Legend: IA kit = immunoassay kit; values are expressed as µg/L; D.L. = detection limit.

From Table 5 it appeared that taking mean values no false-positive values and only one false-negative (M01 carbendazim) were found in the kits. Concentrations above the EC norm could readily be detected. With regard to isoproturon, levels were very low and even below the detection limit of the HPLC method. In this case the sensitivity of the immunoassay is much better. Regression analysis revealed that for the carbendazim immunoassay results were highly correlated with HPLC/DAD (r=0.88397, n=10), whereas no significant correlation could be found for the other kits. Immunoassay results for atrazine were generally lower than those found by GC-MS; and urea-herbicides showed a tendency to be higher than HPLC values.

4 DISCUSSION

Extensive deskresearch has revealed that there are various assays that may be used as indicative screening tools in water quality control to discriminate contaminated from uncontamined samples. However, such assays should meet several criteria such as rapidity, robustness, simplicity, sensitivity and availability of materials and equipment. Three different approaches were selected for evaluation: toxicity testing, HPLC-fingerprinting and immunoassay. From these, toxicity testing showed a lack of sensitivity and to be able to obtain useful results an extensive preconcentration would be required. Together with problems in reliable interpretation of results and the uncertainty of which substance(s) in a sample could have caused the effect observed, makes this kind of assays unsuitable for introduction at unexperienced analytical laboratories. HPLC-fingerprinting and immunoassay were selected as candidate methods for further investigation. Performance testing indicated that both methods did meet the criteria with regard to selectivity, sensitivity, reproducibility, linearity, recovery and matrix effects. The difference between both methods is that HPLC-fingerprinting yields a sum parameter based on the total peak sum (PS) in relation to total detected organic matter (TDOM) and expressed as the ratio PS/TDOM. In contrast, immunoassay results are given as concentration expressed as standard equivalent in µg/L.

Surface water normally contains a varying and often huge amount of humic acids characterized by the corresponding value for TDOM (Table 3). Spiked samples could be discriminated from unspiked samples on the basis of PS and/or the ratio PS/TDOM, with the exception of one sample (PP14D versus PP11D). The advantage of HPLC-fingerprint is the possibility to apply alternative detection methods such as HPLC with diode array detection. This enables further characterization of single peaks when compared to chromatograms of known pesticides with regard to retention times.

Immunoassays allow the detection and optionally quantification of single compounds or a group of related compounds at very low levels[3]. The detection limit is comparable to that of conventional analytical methods or lower, such as in the case of isoproturon. Spiked surface water samples could be readily discriminated from unspiked samples, even in the case of cyclodienes. These pesticides are relatively apolar compounds which adhere to particulate matter. As a result there was found a significant effect of pretreatment of the samples. Moreover, cyclodienes show some background levels in the kit, but these are significantly lower than the values found in spiked samples.

For further validation and implementation of the immunoassay unknown surface water samples were measured in an interlaboratory study by unexperience technicians. The results shown in Table 5 indicate that when pesticides are present, they can be detected by ELISA as well as by conventional methods. Absolute values, however, may be different. For example, the immunoassay results for atrazine are generally lower than found by GC-MS. Partly this may be explained by the use of a high-sensivity kit with a relatively low working range. Using a normal atrazine kit would propably give more reliable results for contamined samples. In contrast, urea-herbicides were found at elevated levels compared to HPLC analysis. This may be due to compounds cross-reacting in the kit. Isoproturon values were too low being below the detection limit of the HPLC method to allow comparison. The performance of the carbendazim kit was such that this ELISA may be used not only for screening purposes, but also for quantification of contamination with this pesticide.

In conclusion, it may be stated that in water quality control both HPLC-fingerprint and immunoassay may be used as indicative methods to assess contamination of surface water. HPLC-fingerprint has the advantage that, besides its discriminating power, in one single run an overview of the number and level of contaminating substances may be obtained, including the possibility of further characterization of single peaks. Disadvantage is the significant contribution of humic acids that may complicate interpretation of results. Immunoassays having similar discriminating power, offer the advantage of rapid multi-sample screening of particular compounds or groups of related compounds at low costs. Some commercially available kits may even be used for quantitative analysis.

References

1. A. Willemsen, M.A. Vaal and D. de Zwart, 'Microbiotests as tools for environmental monitoring, RIVM Report 60742005, 1995.
2. Th.H.M. Noij and A. Brandt, 'Analysis of pesticides in ground and surface water II, H.-J. Stan, Springer-Verlag, Berlin, 1995.
3. E.P. Meulenberg, *Food technol. biotechnol.*, 1997, **35**, 135.

Acknowlegdements: This project was initiated and financed by STOWA, the Netherlands. The porject was coordinated by Prof. Dr. J. Dogterom of the International Center of Water Studies. The investigation was conducted by AquaSense, Amsterdam (Toxicity studies), Kiwa, Nieuwegein (HPLC-fingerprint), and ELTI Support, Nijmegen (ELISAs).

ANALYTICAL MICROSYSTEMS - AN OVERVIEW

D. H. Craston and S. Cowen

LGC (Teddington) Ltd
Queens Road
Teddington
Middlesex TW11 0LY

1 INTRODUCTION

A good analytical laboratory can make reliable and accurate chemical measurements, but these come at a cost and with a time delay from the moment of sampling. In our experience, most decision makers do not like spending money or waiting for data: in fact, in areas like process analysis, clinical diagnostics and pollution monitoring, there are good reasons why results are required quickly and preferably in real time. There is, therefore, without doubt a large market for measurement systems that are small, fast, easy to use and of low cost.

Over the past two decades sensors and biosensors have received much research attention because (on paper, at least) they satisfy the above criteria for rapid measurement. Indeed, there have been some notable commercial successes, for example the glucose pen for diabetics. However, despite some excellent research effort, it is fair to say that the technology has yet to deliver the desired impact, and the majority of analytical work is still performed in dedicated laboratories. Since there have been numerous reports of successful sensors working in research laboratories, it is reasonable to assume that some difficulties have arisen in translating these into products which are robust, reliable and reproducible in real samples or measurement environments.

In parallel with research developments in sensors and related techniques for at-site measurement, various scientific advances have increased the speed and throughput of analysis in the laboratory. Some of these advances have incorporated new measurement approaches; for example using antibodies to selectively detect the desired component (e.g. ELISA). However, there have also been improvements in the more traditional approach to analytical work, where the sample is processed through a series of stages, which extract and purify the desired analytes before separating all of the components present and detecting (and quantifying) each individually. Techniques such as solid-phase extraction and solid-phase microextraction now allow fast recovery and purification of materials with minimal solvent consumption, and provide the capacity to handle many samples in parallel. Further the trend towards smaller column geometries in chromatography and electrophoresis has greatly reduced the time required for final separation and quantification. All of these improvements have been achieved without compromising the reliability and reproducibility of the measurements.

With further improvements in process speed it is conceivable that the time frame for a laboratory measurement will shrink to a level which is commensurate with analysis at-site. However, to take laboratory measurement out into the field will also require step improvements in instrument size, cost and operational complexity in order to comply with the full range of end-user requirements. Analytical microsystems, often referred to as μTAS, offer the potential to achieve all of these things and could therefore provide the next generation of analytical products.

2 ANALYTICAL MICROSYSTEMS

The research community has yet to define analytical microsystems properly. However, we associate the expression as implying the miniaturisation by microengineering of any measurement process, whether this relates to established technology like chromatography and electrophoresis, or the newer sensor systems. The link with microengineering is an important one, as this is synonymous with high accuracy and resolution, and low cost mass production.

2.1 Microengineering

Microengineering,[1] or microfabrication (sometimes referred to as MEMS, or microelectromechanical systems) has developed out of the microelectronics industry and uses the same basic tools of deposition, lithography and etching to produce topographic features on the surface of substrate materials. There are many mechanisms by which structures can be microengineered,[2] and this can be illustrated simply by the reproduction of a pattern defined in a lithographic mask as an etched feature on the surface of a wafer. This is shown in Figure 1 for a silicon substrate, and demonstrates how the dimensions of the structures so produced can be influenced by the material, and the time and mechanism of etching.

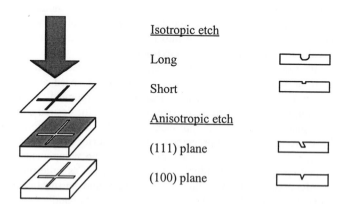

Figure 1 *Production of channels on a silicon substrate by a simple lithography process. A photoresist-covered substrate is exposed to UV light through a mask, the resist developed and the silicon etched. Depending on the etching conditions and wafer orientation, different channel profiles can be obtained*

Typically such structures can be produced over an area of several square centimetres, with a resolution of 1-100 μm both parallel with, and perpendicular to, the plane of the substrate. More complex structures can be prepared by the addition of layers of other materials onto the substrate, and patterning and etching these by successive lithographic steps. These repeated procedures create pseudo three-dimensional surface features. For example, by completely removing some of the deposited underlayers it is possible to produce free-standing structures like cantilevers and membranes which can be made to move under the influence of a suitable force.

Given its roots, it is not surprising that much of the early work in microengineering used silicon as the substrate, and utilised some of the unique physical properties of this material and its crystal structure to produce complex devices that incorporated moving parts (micromachines). This early work has led to several commercially available devices, most notably miniature pressure sensors and accelerometers: the latter are mass produced for use as the sensing elements for actuating airbags in cars. Although silicon is still a favourite material of microengineers, the use of glasses, ceramics and polymers is becoming more prevalent for reasons of both cost and compatibility. As such, these materials are likely to be more prominent in future commercial devices.

The interface between microengineering and chemistry is an obvious one. Most analytical procedures are performed with instruments - microengineering is a proven tool for fabricating instrument components, and is equally well-suited for miniaturising the detection elements of most sensor systems (e.g. electrodes, optical waveguides, FETs and piezo-devices). Prototype micropumps, microvalves and micro-flowmeters can all be obtained from commercial sources. These components could potentially provide the backbone of micro-fluid handling systems such as those used in chromatography, immunoassay and flow injection analysis. The questions to ask therefore are why would you wish to build miniature analytical instruments, what are the likely problems associated with them, and when are you likely to be able to buy them off-the-shelf?

2.2 Analytical Microsystems: The Advantages

The most obvious advantage of analytical microsystems is small size and the portability this brings for application at site, within restricted domains or even in the laboratory where small sample volumes need to be handled. Of equal importance, however, is the improvement in speed of operation achieved when chemical processes are carried out within restricted volumes. This advantage stems largely from the shorter distances for molecular diffusion and for the transfer of heat.

In the context of chemical diffusion, it has been demonstrated that the mixing of reaction streams in microengineered devices is extremely efficient.[3] This has generated significant interest from synthetic chemists, and affords the potential for rapid pre- and post- column derivatisation for chemical measurement. Also of relevance to chemical measurement is the well-established principle that in open-tubular chromatography the resolution and/or speed of separation are enhanced by reducing the diameter of the capillary. Using microengineering techniques we have been able to generate columns for open-tubular liquid chromatography where the theoretical efficiency is similar to standard HPLC (up to 10^4 plates), but the run times are less than one minute.[4] This can be achieved while still retaining reasonable measurement sensitivity.

Improved thermal transfer has a number of benefits in chemical synthesis which includes the potential for energy savings. Heat input and output are also key to a range of analytical processes. For example, thermal cycling in PCR is a limiting time factor in

DNA analysis, the speed of which has been shown to improve significantly when performed in a microengineered reaction chamber. Perhaps the most graphic demonstration of the benefits of improved heat transfer, however, comes from the studies of electrophoretic processes in microengineered systems. Here, high-resolution separations have been achieved in seconds because efficient heat dissipation has allowed the application of very large voltage gradients.

Further benefits associated with analytical microsystems are the potential for scale-up in production at relatively low cost, and the ability to organise structures on the surface of the substrate to maximise space utilisation. It is as simple to lay down an array of closely spaced microchannels on a substrate wafer as it is to produce a single channel. Thus, in principle, a device performing several tens of electrophoretic separations should be as easy and cheap to produce as a single sample device. On a similar note, the scale-up in numbers of wells on a standard microtitre plate to 1536 or more (compare this with the standard 96-well plate) has been demonstrated using microengineering technology. It has also been shown that many centimetres or metres of capillary column can be organised on the surface of a substrate which is only a square centimetre in area using loop or spiral geometries (Figure 2).

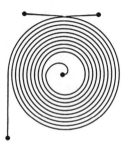

Figure 2 *On-chip column layout for open-tubular liquid chromatography*

This facility to organise features on surfaces could have far reaching benefits in assisting the integration of separate analytical components. An example of this could be in capillary separations, where the joining of capillaries is difficult on a macro-scale but trivial in an on-chip format. This might usefully be exploited in setting up gradient elution systems for chromatography or capillary electrochromatography.

2.3 Analytical Microsystems: the Problems and Challenges

Although individual components of an analytical microsystem have been described at a research level, there are no portable measurement systems on the market that are fabricated entirely by microengineering processes. While the level of commercial investment in the area would suggest that the first products will soon be released, the technical issues that need to be bridged should not be underestimated. Main issues include scale-up to low-cost production, component integration and materials compatibility, and the interface with real samples, including the sampling process.

Mechanisms for component integration are particularly important. The following section illustrates that individual analytical processes have been replicated in miniature format, and shown to work, albeit in a research environment. However, an analytical microsystem is likely to require a number of these components combined in a suitable configuration. The skill therefore will be in bringing devices together without affecting the overall system performance, as could occur for example by introducing significant dead volume. Given that the structures in a complete analytical microsystem might typically contain a total volume of a few tens of microlitres, it is easy to see that the task of integration is not a trivial one.

The low volume of analytical microsystems also implies that the amount of sample analysed in a single run will be extremely small. Consequently there are issues around sampling and the need to obtain information that properly represents the complete measurement environment. The size of the channels and chambers of an analytical microsystem also means that the potential for blockage by particulates present within the analytical solutions or sample is high. The interface with the sample is extremely important, and for many applications will determine the viability of commercial systems. The success of current research into components for sample preparation is therefore considered pivotal in defining the future scope of the technology.

3 CURRENT RESEARCH THEMES

Over the last few years, there has been a rapid increase in the diversity of research and number of groups working in this field, and the literature now contains examples of miniaturisation applied to nearly all aspects of chemical analysis. There is now an established regular international symposium devoted to analytical microsystems[5-7] which continues to grow steadily, and considerable investment is being attracted for the development of commercial systems. Some key areas are briefly reviewed below.

4.1 Microfluidic handling

The control of fluid flow and material transport is fundamental to chemical analysis, and understanding how to do this in a microchip device is crucial to the development of complete, miniaturised measurement systems. A typical analytical process will involve sample intake, movement and switching of flow streams, as well as mixing and reaction steps before the final measurement is generated. On the micro-scale, the associated physical and chemical characteristics are often quite different from those observed with benchtop instruments, and studies of microreactors[8] and mixing[3] have demonstrated the advantages of reduced dimensions. Miniature versions of pumps and valves have also been constructed,[9,10] and their performance is improving, but at present their capabilities are limited in terms of the pressure delivery available and robustness (lifetime).

Alternatives to pumps have therefore been sought, and electrokinetic methods have proved to be very successful in generating flow without moving parts. Capillary electrophoresis (CE) is a well-known technique for moving ions in a highly controllable manner under the influence of an electric field, while the combined effect of the applied field and the electrical double layer present at the liquid/channel wall interface moves neutral species by viscous drag of the solvent - a process known as electroosmosis. This can be achieved with a field strength of less than 1 kV cm^{-1} and so with careful selection of conditions, solutions can be moved and samples injected with relative ease and low power consumption. It is also possible to use electric fields for the manipulation of bioparticles such as cells and bacteria by dielectrophoresis. In this way the separation of cells in blood has been demonstrated,[11] as has the trapping of particles in electric field cages.[12]

4.2 Separation Science

The first application of microengineering to chemical analysis was published in 1979, when Terry and co-workers described a gas chromatograph fabricated on a silicon wafer.[13] Containing a micromachined injector, column and thermal conductivity detector, the system was able to separate a three-component sample in less than 10 seconds, although the resolution was not as high as that of a conventional system. However, despite the promise shown by the microengineering approach, and the improved performance offered by small-bore columns,[14] it was only at the end of the 1980s that microchip devices were again considered as a way of obtaining rapid separations.[15,16] Using capillary electrophoresis (CE) devices fabricated on glass chips, separations of high resolution were obtained in only a few minutes. The relative simplicity of the technique, together with the fact that sample injection is achieved with a completely valveless design, has meant that CE on chip has since become very well-characterised, and the early work followed with developments such as interfacing with mass spectrometry[17] and sub-millisecond separations.[18] More recently this has been extended to capillary electrochromatography,[19] with both packed and open-tubular columns.

Miniaturising liquid chromatography however, is technically more difficult, since the mobile phase and sample are pressure-driven. For open-tubular columns the diameter required for an acceptable resolution is of the order of 1-2μm,[14] requiring pumps of high capacity which are not available yet; while packing presents considerable technical difficulties. Nevertheless, there has been some progress in this area,[20,21] and it is expected that this will continue.

4.3 DNA Analysis

The polymerase chain reaction (PCR) is now established as the chosen route for DNA analysis, and has found a place in medicine, and the food, environmental and forensic sciences. As a thermal cycling technique, it lends itself very well to miniaturisation, and there is huge commercial potential for PCR devices offering the enhanced performance obtainable. Examples of microfabricated reaction vessels with integrated heaters have been described,[22] and these show that the speed of PCR can be increased such that around 35 cycles can be carried out in a few minutes. Recently, a portable instrument for DNA analysis in the field, containing a PCR chamber coupled to a microfabricated CE column was reported.[23] DNA diagnostics is thought to offer an enormous market for chip devices, linked as it is with the Human Genome Programme, and drug development. Commercial activity is currently intense, and includes not only work on PCR but also on high-density sensor array devices[24] for detecting large panels of DNA sequences. With advances in sensitive detection the potential could exist for diagnosis without prior amplification.

4.4 Biochemical analysis

The trend towards greater use of biochemical tests in the laboratory (largely for screening) and in diagnostics is being mirrored in current developments in chip technology, and devices performing biochemical analyses have been described.[25,26] Most reported examples have considered the same approach currently adopted in the macro-scale or in sensor development, and used microfabrication to increase substantially the numbers of measurements possible on a single device. Much of the current effort in this area has been catalysed by the pharmaceutical industry in their quest for higher throughput to cope with the analytical and screening demands arising from the adoption of combinatorial chemistry in drug discovery. This, combined with the significant and expanding market in diagnostics, has made chip bioassays a further hotbed of commercial activity.

Manipulation, counting and sorting of single cells is a further interesting application of analytical microsystems. The tools for these include dielectrophoresis (mentioned previously) and flow cytometry.[27] Intricate measurements on single cells have been reported,[28] demonstrating the potential for chip technology as a tool for leading-edge research and development.

4 CONCLUSIONS

The viability of fabricating and operating analytical microcomponents offering rapid measurement has now been well-documented. The volume of research in this area continues to grow and generate significant commercial interest, particularly in DNA and clinical diagnostics, and combinatorial chemistry. If the technical issues, such as component integration and real-world sampling can be successfully tackled, then analytical microsystems will provide another route for achieving reliable measurement outside the laboratory.

References

1. R. F. Wolffenbuttel, *Sensors and Actuators*, 1992, **A30**, 109.
2. 'Sensors: A Comprehensive Survey', Vol. 1, T. Grandke and W. H. Ko (eds), VCH Press, Weinheim, 1989.
3. R. Miyake, T. S. J. Lammerink, M. Elwenspoek and J. H. J. Fluitman, *Proc. MEMS Workshop 1993*, p248.
4. S. Cowen and D. H. Craston, in ref. 5, p295.
5. 'Micro Total Analysis Systems', A. van den Berg and P. Bergveld (eds), Kluwer, Dordrecht, 1995.
6. Proceedings of the 2nd International Symposium on Miniaturised Total Analysis Systems µTAS96, *Analytical Methods and Instrumentation* (Special Issue), 1996.
7. 'Micro Total Analysis Systems 98', D. J. Harrison and A. van den Berg (eds), Kluwer, Dordrecht, 1998.
8. 'Microsystem Technology for Chemical and Biological Reactors', DECHEMA, **132**, Mainz, 1996.
9. M. Esashi, S. Shoji and A. Nakano, *Sensors and Actuators*, 1989, **20**, 163.
10. B. H. van der Schoot, S. Jeanneret, A. van den Berg and N. F. de Rooij, *Sensors and Actuators*, 1992, **B6**, 57.
11. G. H. Markx and R. Pethig, *Biotechnol. Bioeng.*, 1995, **45 (4)**, 337.
12. G. Fuhr in ref. 6, p39.
13. S. C. Terry, J. H. Jerman and J. B. Angell, *IEEE Trans. Electron Devices*, 1979, **ED-26**, 1880.
14. J. W. Jorgenson and E. J. Guthrie, *J. Chromatogr.*, 1983, **255**, 335.
15. A. Manz, N. Graber and H. M. Widmer, *Sensors and Actuators*, 1990, **B1**, 244.
16. A. Manz, D. J. Harrison, E. Verpoorte and H. M. Widmer, *Adv. Chromatogr.*, 1993, **33**, 1.

17. Q. Xue, F. Foret, Y. M. Dunayevskiy, P. M. Zavracky, N. E. McGruer and B. L. Karger, *Anal. Chem.*, 1997, **69**, 426.
18. S. C. Jacobson, C. T. Culbertson, J. E. Daler and J. M .Ramsey, *Anal. Chem.*, 1998, **70**, 3476.
19. S. C. Jacobson, R. Hergenröder, L. B. Koutny and J. M. Ramsey, *Anal. Chem.*, 1994, **66**, 2369.
20. G. Ocvirk, E. Verpoorte, A. Manz, M. Grasserbauer and H. M. Widmer, *Analytical Methods and Instrumentation*, 1995, **2**, 74.
21. B. He and F. E. Regnier, *Anal. Chem.*, 1998, **70**, 3790.
22. A. T. Woolley, D. Hadley, P. Landre, A. J. de Mello, R. A. Mathies and M. A. Northrup, *Anal. Chem.*, 1996, **68**, 4081.
23. M. A. Northrup, W. Bennett, D. Hadley, P. Landre, S. Lehew, J. Richards and P. Stratton, *Anal. Chem.*, 1998, **70**, 918.
24. See for example www.affymetrix.com.
25. N. Chiem and D. J. Harrison, *Anal. Chem.* 1997, **69**, 373.
26. D. J. Harrison, K. Fluri, N. Chiem, T. Tang and Z. Fan, *Technical Digest, Transducers 95*, 8th International Conference on Solid-State Sensors and Actuators, Stockholm, June 1995, p752.
27. A. Wolff, U. D. Larsen, G. Blankenstein, J. Philip and P. Telleman in ref. 7, p77.
28. C. D. T. Bratten, P. H. Cobbold and J. M. Cooper, *Anal. Chem.*, 1997, **69**, 253.

RAPID DETECTION OF CHANGES IN FLUID COMPOSITION USING ON-LINE IMPEDANCE SPECTROSCOPY

M. E. H. Amrani, R. M. Dowdeswell and P. A. Payne

Department of Instrumentation and Analytical Science
UMIST
Manchester M60 1QD

1 INTRODUCTION

Conductivity measurements were among the first of the instrumental methods to be developed for studying properties of electrolyte solutions. Conductivity is proportional to the total ionic content of a solution and, therefore, is non-specific. This lack of specificity limits the use of conductivity measurements to applications where either only a single electrolyte is present or where the total ionic content needs to be measured. Over the years, conductivity cells have been developed which provide remarkable sensitivity and, more recently, high frequency conductivity measurements have been developed which enable changes in conductance or dielectric constant to be measured without the introduction of electrodes into direct contact with the solution.

The electrical properties of a material held between two electrodes separated by a fixed distance are given by $G = \dfrac{\sigma A}{d}$ and $C = \dfrac{\varepsilon_0 \varepsilon A}{d}$ where A is the electrode area, d the separation, G the conductance, C the capacitance, σ is the electrical conductivity (Siemens per metre), ε_0 is the dielectric permittivity of free space, and ε is the relative permittivity of the material separating the electrodes. These parameters completely characterise the material and, therefore, measurements of this nature can provide vital data of a fundamental scientific nature, but they also lend themselves to extensive use in the process industries. In the majority of applications where high frequency measurements are made, these are usually limited to a single frequency. In our laboratory over the past year or so, we have studied the use of high frequency dielectric measurements in which a range of measurement frequencies are employed. We have also combined this with the use of the resonance effect in which the conductance cell or, more properly, dielectric cell, is made to form part of a resonant circuit. This enables us to track small changes in dielectric constant of the material between the electrodes, giving rise to measurements of contaminants in pure water, for example, in which the detection limits are down to a few parts per billion (ppb).

2 FLOW CELL AND INSTRUMENTATION

Details of the flow cell have been given elsewhere[1]. The construction is extremely simple comprising a pair of electrodes on the outside of a pipe made of an insulating material such as perspex. The whole structure is encased in a metal shielding in order to prevent extraneous noise signals being picked up. This also has the added bonus of preventing radiation from the flow cell affecting other instruments. The two electrodes are connected to the input terminals of a Hewlett Packard impedance analyser (HP4191A) which can be set to measure one of a number of electrical parameters as a function of frequency. In many instances, we choose to study dissipation factor as a function of frequency where dissipation factor is simply the real part divided by the imaginary part of the flow cell impedance. As mentioned previously, we operate the flow cell as part of a resonant circuit. This is achieved simply by choosing an appropriate inductor connected in series with the flow cell. It proves convenient to arrange the inductor such that the whole structure is resonant at around 1 MHz, although as we show later, the resonant frequency can vary quite widely, dependent on what is flowing through the cell.

3 LOW LEVEL CONTAMINANTS IN WATER

We conducted a series of experiments in which distilled water was allowed to flow through the cell at a rate of 1 litre per minute in a closed loop configuration. The distilled water dissipation factor was then measured and a series of repeat measurements were made showing that the repeatability of the frequency at the peak value of the dissipation factor was 5 Hz. As can be seen in Figure 1, when one part per billion of commercial fertiliser was added to the distilled water, the dissipation shift was some 60 Hz, ie, twelve times the repeatability figure. As more fertiliser was added, so the dissipation factor moved progressively up the frequency scale. Repeatability of the peak value of the dissipation factor peak is prone to some variation since the computation, ie, resistance divided by reactance becomes rather uncertain as the reactance approaches zero.

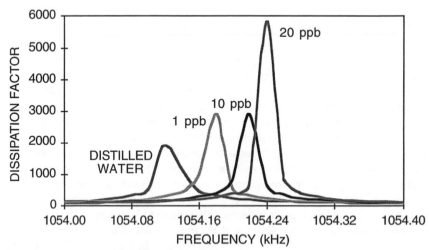

Figure 1 *Effect of fertiliser on water*

4 APPLICATIONS IN THE FOOD AREA

The measurement technique is equally applicable to foodstuffs and as an example we have examined the response of the flow cell to two different types of olive oil. These were both extra virgin olive oils, one from Morocco and the other from Tunisia. The results of these measurements are shown in Figure 2. As can be seen, the resonant frequencies are much lower than previously, but the difference between the two sets of data is some 15.3 kHz, vastly greater than the repeatability value established previously. Each of the oils were sampled and measured five times to produce the data shown in Figure 2.

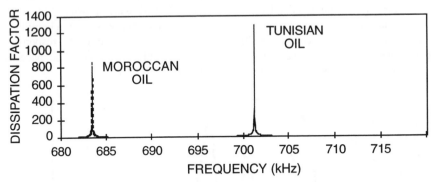

Figure 2 *Dissipation factor as a function of frequency for two olive oils*

As a further example, we have used the flow cell to measure the difference between two colas. These data are shown in Figure 3, where the frequency difference is seen to be some 50 Hz, again, well above the repeatability figure.

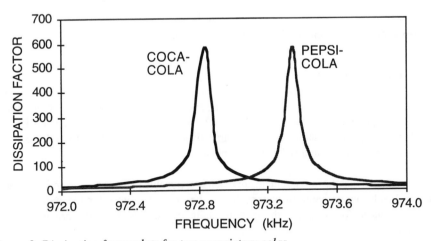

Figure 3 *Dissipation factor plots for two proprietary colas*

We have also examined the use of the flow cell in the context of monitoring a product such as beer for the presence of 1 ppm glycol contamination. Our results confirm that we can detect such contaminants down to levels well below the accepted figure in the brewing industry[2].

5 DYNAMIC RESPONSE

An alternative way of using the instrument is to set the impedance analyser at a fixed frequency near the resonant frequency for the cell and then to monitor changes in one of the electrical parameters such as the reactance. We have conducted experiments of this nature on the flow cell in which tap water is allowed to flow through and to which boluses of water contaminated by corroding metal have been added. These data are shown in Figure 4 for two injections of contaminated water. As can be seen, the response is extremely rapid and the recovery to baseline occurs within 20 or 30 seconds. The dynamics of this response are largely governed by the volume of water in the cell and obviously this could be made very much smaller and the dynamic response would then be even faster.

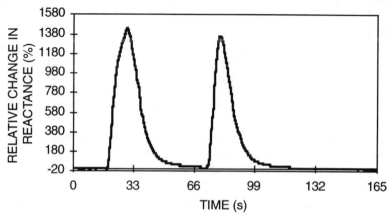

Figure 4 *Reactance changes for injections of contaminated water*

6 CONCLUSIONS

In this paper we have demonstrated the performance of a flow cell used as part of a resonant impedance spectrometer. At present the system is based on the use of a Hewlett Packard impedance analyser, but alternative ways of impedance measurements are under development which are not only potentially lower cost, but also faster.

We have shown that the measurement approach is applicable to the detection and monitoring of low level contaminants in water. It also has potential uses for the monitoring of contaminants in foodstuffs, in the brewing industry and possibly for lubricants used in machinery.

References

1. P.A. Payne, M.E.H. Amrani and R.M. Dowdeswell, 'The use of on-line impedance spectroscopy for the low level detection of fluid contamination', Paper to be presented at Sensor 99 Conference, Nurenberg, Germany, 18th May 1999.
2. R.S. Lewis (editor), 'Sax's Dangerous Properties of Industrial Materials', 8th edition, Van Nostrand Reinhold, New York, 1992, Vol. 2, p. 1219.

THE ELECTROCHEMICAL DETECTION OF PENTACHLOROPHENOL (PCP) USING THE INHIBITION OF LACTATE DEHYDROGENASE (LDH) AS A MODEL SYSTEM

S. J. Young, A. A. Dowman, J. Hart & D. C. Cowell

Faculty of Applied Sciences
University of the West of England
Frenchay Campus
Coldharbour Lane
Bristol   BS16 1QY

1 INTRODUCTION

Pentachlorophenol (PCP) has been used extensively as a multi-purpose biocide for many decades, particularly in the wood-preservation industry.[1] It is included in the European Union directive 80/778/EEC which provides statutory guidelines for maximum admissible concentrations (MAC) of undesirable compounds in drinking water. The MAC for PCP is 0.1 µg l^{-1} (3.75×10^{-10} M). The World Health Organisation recommends a maximum PCP residue limit of 10 µg l^{-1} based on toxicological thresholds.[2]

PCP may be detected using electrochemical enzyme-inhibition biosensors[3] where reduction of an enzymatic reaction would be detected by a fall in transducer output. Inhibition of the enzyme lactate dehydrogenase (LDH) by PCP was shown to be increased using a non-optimal enzyme cofactor, NADPH.[4] Inhibition by PCP increased with decreasing NADPH concentration (a characteristic of competitive inhibition) but detection limits were constrained by the insensitivity of the spectrophotometric detection method, where the lowest NADPH concentration that produced a measurable reaction was 20 µM NADPH.

This study aims to perform inhibition assays using a modified screen printed carbon electrode (SPCE) combined with the sensitive technique of amperometry in stirred solution.[5] This involves the linking of LDH with another enzyme, lactate oxidase (LOD) to produce a measurable electroactive species, H_2O_2. The move from spectrophotometric to electrochemical detection would provide an initial step in the production of biosensors for environmental diagnostics. It should be added that an assay based on an electrochemical sensor offers the advantage, over the spectrophotometric assay, that determinations can be performed on opaque water samples

2 MATERIALS AND METHODS

2.1 Reagents

Rabbit muscle LDH (*EC. 1.1.1.27*) was purchased from Biozyme (Blaenavon, UK) at a specific activity of 459 U mg^{-1} protein. LOD, from *P. aureas*, (*EC. 1.1.3.2*) at a specific activity of 182 U mg^{-1} protein, was obtained from Genzyme (West Malling, UK). All reactants were of analytical grade from Sigma Chemical Co. (Poole, UK). PCP (>99 %) was dissolved in one part ethanol and mixed with 24 parts 50 mM potassium phosphate buffer (pH 7.4) to give an initial assay concentration of 70 µM.

2.2 Amperometry in Stirred Solution

Modified screen-printed carbon electrodes (SPCEs)[5] containing five percent cobalt phthalocyanine (CoPC) and a 2 % drop-coated cellulose acetate membrane were used to detect H_2O_2 from a linked enzyme system containing LDH and LOD (Scheme 1). The SPCE was connected to a BAS LC-4B amperometric detector and ABB SE120 chart recorder to monitor the change in current (nA min^{-1}).

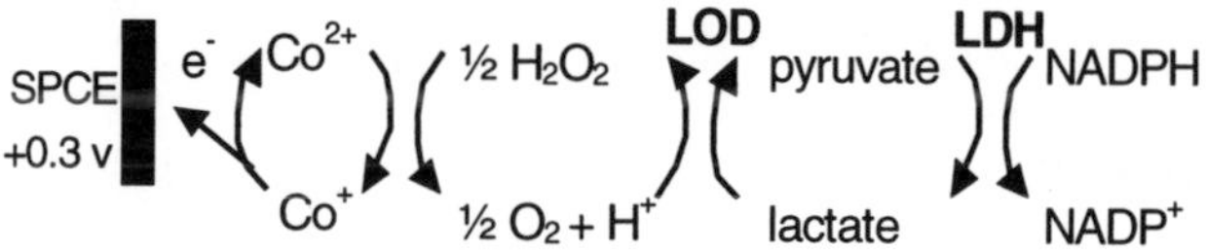

Scheme 1

All assays were carried out in 50 mM potassium phosphate buffer, pH 7.4, at 25 ± 0.1 °C in a total volume of 5 cm^3. LOD assays were modified from published data[6] and were analysed for inhibition by 70 µM PCP under rate-limiting conditions in the absence of the linked LDH assay. Linked assays, initiated by the addition of 20 µM NADPH (the rate-limiting step) in the presence of 800 µM pyruvate, were modified with respect to LOD :LDH ratio. PCP detection limits were obtained using assay conditions shown to provide the greatest level of inhibition.

3 RESULTS

3.1 LOD Inhibition

Figure 1 shows Lineweaver-Burke plots for lactate in the presence and absence of 70 µM PCP. The intersection on the y-axis was indicative of competitive inhibition. Kinetic parameters were calculated using Enzpack computer software (Biosoft, UK). The Michaelis-Menten constant, K_m, for lactate was 159 µM and the PCP inhibition constant, K_i, was 278 µM. 70 µM PCP caused 19.3 ± 1.7 % (mean ± SEM, n =12) inhibition of LOD activity using 20 µM lactate and 0.006 U/ml LOD in the assay. LOD activity was found to be rate-limiting below 1U/ml when 20 µM lactate was used.

3.2 Linked Assay Modification

The activity of LDH was required to be greater than that of LOD in order to produce a measurable reaction velocity. The ratio of LDH: LOD selected was initially 5: 0.8 U/ml (all rate-limiting). The activity of LDH corresponded to that chosen in previous studies where 63 % inhibition of LDH activity was observed.[4] The activity of LOD was higher than that used in the LOD inhibition study in order to ensure a significant rate of reaction in the linked system.

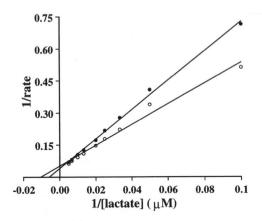

Figure 1 *Lineweaver-Burke plots for LOD with (●) and without (○) 70 μM PCP*

3.3 Linked Assay Inhibition by PCP

The linked assay system was inhibited by 45.7 ± 1.1 % (mean ± SEM, n =17) by 70 μM PCP. Figure 2 shows inhibition of the linked bi-enzyme system by various concentrations of PCP using 5 U/ml LDH and either 0.8 or 0.1 U/ml LOD. There was no difference in PCP detection between the two assays. In each case the calibration was curved with a PCP detection limit (EC_{10}) of 10 μM.

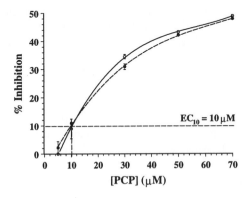

Figure 2 *PCP detection with LDH: LOD at 5: 0.8 U/ml (○) and 5: 0.1 U/ml (●)*

4 DISCUSSION

4.1 LOD Inhibition

PCP competitively inhibited LOD with a K_i of 278 µM, which is far in excess of the initial PCP concentration of 70 µM. Therefore, significant inhibition of LOD would not be expected, as shown by the level of 19.3 % inhibition using 70 µM PCP. This amount of inhibition was observed using an assay concentration of 20 µM lactate and 0.006 U/ml LOD. 20 µM lactate was chosen since this is the concentration that would be generated in the linked assay upon introduction of 20 µM (rate-limiting) NADPH.

4.2 Linked Assay Inhibition

The overall inhibition of H_2O_2 production from the linked assay was less than that obtained spectrophotometrically with LDH.[4] The calibration plots were also different, with the linked system producing a curved graph, with an EC_{10} of 10 µM, and the LDH assay[5] producing a linear graph with an EC_{10} of 1 µM. Reduced inhibition may be attributable to the fact that both enzymes are competitively inhibited by PCP and will bind PCP. Binding of PCP to LOD will reduce the amount available to inhibit LDH, which exhibits greater inhibition characteristics.

Reduction of LOD activity may help to reduce the unwanted effect of "PCP-sequestering". However, reduction from 0.8 to 0.1 U/ml LOD did not improve inhibition data. LOD data showed that activities below 1 U/ml were rate-limiting for assays containing 20 µM lactate, indicating that chosen LOD activities were all rate-limiting. However, when the activity chosen for the LOD inhibition assays is compared (0.006 U/ml) it can be seen that there is scope for further reduction.

Once levels of inhibition have been increased to that obtained previously using LDH[4], NADPH can be reduced to the lowest possible concentrations. This should increase competitive inhibition by PCP and enable lower detection limits to be achieved. The activity of LDH would have to be modified to ensure that it was still rate-limiting with respect to the chosen NADPH concentration. Sensitive electrochemical detection of PCP using inhibition of LDH could be used as a model system to develop electrochemical biosensors to detect other aquatic pollutants of concern.

References

1. G. Renner and W. Mücke, *Toxicol. Environ. Chem.*, 1986, **11**, 9.
2. World Health Organization "Guidelines for Drinking Water Quality", Geneva, 1984.
3. D. C. Cowell, A. A. Dowman and T. Ashcroft, *Biosensors & Bioelectronics*, 1995, **10**, 509.
4. S. J. Young, A. A. Dowman and D. C. Cowell, *J. Pesticide Biochem. & Physiol.*, 1999, *In press*.
5. J. P. Hart and S. A. Wring, *Electroanalysis*, 1994, **6**, 617.
6. Lockridge, V. Massey and P. A. Sullivan, *J. Biol. Chem.*, 1972, **247**, 8097.

MICROBIOTESTS FOR RAPID AND COST-EFFECTIVE HAZARD ASSESSMENT OF INDUSTRIAL PRODUCTS, EFFLUENTS, WASTES, WASTE LEACHATES AND GROUNDWATERS

Prof.em.Dr.G.Persoone

Laboratory for Environmental Toxicology and Aquatic Ecology, Ghent University, Ghent/Belgium

1 INTRODUCTION

Until rather recently it was thought that the potential impact of industrial products and/or their wastes released into the environment could be determined by "exclusive" chemical investigations, i.e. by the qualitative and quantitative analysis of the individual compounds present in the products. The hazard of the "multi-chemical" products was then extrapolated either from the most toxic compound, or by summing up the toxicities of the individual chemicals.

To date it is admitted that the former approach does in most cases not give a correct estimate of the "real hazard" of products composed of mixtures, and this for the following reasons :

a) the qualitative and quantitative analysis of "all" the individual chemicals which can be present in (mixed) products and/or wastes constitutes not only a serious technical problem, but also a substantial financial burden, chemical analyses of mixtures is in practice very often limited to a restricted number of compounds.

b) even with products or wastes containing only a few chemicals, the ultimate toxicity is not the sum of the individual toxicities. Phenomena such as e.g. bioavailability, synergistic or antagonistic effects indeed always dictate the ultimate hazard of the mixtures.

The only way to get an "integrated" answer to the hazard of multi-chemical products and wastes is to expose organisms of the receiving environment to these products or wastes in so-called "bioassays". The effects of the chemical mixtures in the toxicity tests are measured in terms of mortality, growth inhibition or inhibition of reproduction, depending of the time of exposure, the concentration of the chemicals and their intrinsic toxicity.

The biological part of the "receiving environment" (aquatic or terrestrial) has a well-defined structure and function, and all the biota have a specific role in the biological chain of "production-consumption-decomposition" which ensures the normal functioning and hence the health of the ecosystems of which man is ultimately totally dependent.

Ecotoxicological evaluations with only one test species only reveal how toxic a single chemical or a multi-chemical sample is to that species. A single test can, however, not tell whether other important groups of biota may possibly even be more affected by exposure to the same chemical(s). Consequently there is a real danger, when using only one bioassay

- with whatever test species - to underestimate the real hazard of the chemical or the product analysed.

Toxic effects are indeed "species specific" as well as "chemical specific". It is thus necessary, when attempting to estimate the hazard of compounds or products released in the environment - and this is automatically the "ultimate fate" of all man-made products - to use a battery of tests with species representative of the different links in the trophic chains.

Bioassay methodologies have been worked out over the last 30 years with selected test biota, for the determination of the intrinsic toxicity of pure chemicals. A limited number of these tests have been endorsed in the meantime by international organizations such as e.g. the OECD and the ISO.

2 PROBLEMS WITH CONVENTIONAL TOXICITY TESTS

The application of "conventional" toxicity tests is seriously hampered by the need for the continuous culturing and/or maintenance of live stocks of the test species, in good health and in sufficient numbers, and the high costs associated with this.

A study was made by Persoone and Van de Vel in 1987 to evaluate the costs of the 3 acute and the 2 chronic aquatic tests prescribed by the OECD in its "Guidelines for the Testing of Chemicals". This study was based on questionnaires sent out to European laboratories involved in toxicity testing, with regard to the costs of equipment, consumables and personnel involved in the performance of the bioassays. The answers received from about 40 laboratories revealed that all the "conventional" ecotoxicological tests are quite expensive, i.e. cost on the average more than 500 Euro up to more than 1000 Euro for acute bioassays, and several thousand Euro for the chronic toxicity tests.

Looking at the breakdown of the costs in percentage of equipment, consumables and personnel, it further appeared from the answers received that 90% of the costs come from the category "personnel", and that 50% of the latter costs are not related to the performance of the toxicity tests, but to the year-round culturing and maintenance of the live stocks of the test species.

A major consequence of the necessity for continuous culturing/maintenance of live test stocks and the high costs associated with this, is that ecotoxicological testing is to date, in most countries, limited to a small number of highly specialized laboratories. Testing furthermore is restricted quantitatively as much as possible.

From the more than 60.000 pure chemicals which are produced and used worldwide, only a very small part has been analysed so far for their "intrinsic toxicity", i.e. the hazard which they may constitute for living organisms upon their (deliberate or accidental) release into the environment.

In 1996 the European Chemical Bureau estimated that even for the 2500 HPVC's (High Priority Volume Chemicals), i.e. compounds produced in very high tonnages, only 55% had been analysed for their acute toxicity to fish or Daphnia, only 20-30% for their short term ecotoxicity to algae, and only 5% for their toxicity to terrestrial organisms.

Over the last decades legislation has become much more strict for the introduction of "new" chemicals, and every producer of a new compound has to submit a "notification dossier" to the competent authorities to obtain permission to release the compound on the market. Besides information on the toxicity of the chemical to humans, the dossier also has to provide data on the toxicity to aquatic and terrestrial biota.

As can be extrapolated from what is said above, the costs related to the toxicity identification of each new chemical (of which more than 1.000 are put yearly on the market) are also very substantial. Each bioassay performed indeed is dependent again on the availability and hence the culturing and maintenance costs of the test species needed to perform the prescribed tests .

3 CULTURE/MAINTENANCE FREE MICROBIOTESTS AS ALTERNATIVES TO CONVENTIONAL TOXICITY TESTS

The awareness of the biological, technical and not the least the financial burdens associated with "conventional" bioassays and the increasing demand for tests which can be used for "routine" analysis of (multi-chemical) products and their wastes, has automatically triggered the development of alternative assays. One of the first approaches was to try to miniaturise the bioassay technologies by using mainly unicellular or small multicellular organisms, hence the name small-scale tests or "microbiotests ".

According to Blaise (1991), one of the "missionaries" for alternative bioassays, a microbiotest should have the following characteristics :
1. inexpensive or cost-effective
2. generally not labour intensive
3. have a high sample throughput potential
4. cultures that are easily maintained or maintenance free
5. modest laboratory and incubation space requirement
6. low costs of consumables (e.g. disposable test containers)
7. low sample volume requirements

These prerequisites should, however, not lead to loss of precision, repeatability nor sensitivity in comparison to conventional bioassays.

An excellent overview of bioassays with smaller life forms, enzyme systems, tissue cultures and young life forms was recently compiled by Wells *et al.* (1998) in the book « Microscale Testing in Aquatic Toxicology ». The large majority of the small-scale assays described, however and unfortunately, still relies on (continuous) culturing or maintenance of live stocks of test species !

Until recently only 2 internationally accepted microbiotests could claim for full independence from stock culturing, namely the luminescence inhibition assay with lyophilised marine bacteria (Bulich, 1979), commercially available under the name Microtox® and Lumistox®, and the seed germination/growth test with plant seeds (OECD, 1984).

Extensive research performed during the last 15 years in the Laboratory for Biological Research in Aquatic Pollution (LABRAP) at the Ghent University in Belgium (presently renamed Laboratory for Environmental Toxicology and Aquatic Ecology), has fortunately led to the development of additional culturing/maintenance free microbiotests using various test species from different phylogenetic groups. These small-scale bioassays were given the generic name « Toxkits » (Persoone, 1991). Toxkit microbiotests can be performed anywhere- anytime since they depart from biological material in dormant (cryptobiotic) or immobilised form, with a shelf life of several months up to several years. Adhering to the miniaturisation prerequisites of small-scale assays described by Blaise (1991), the Toxkits, besides the test biota, also contain all the materials needed for the performance of the assays, such as e.g. concentrated media, small test containers, micropipettes, etc.). This makes their application in practice simple, practical and cost-effective. The precision and the repeatability of these assays is also better than that of conventional ecotoxicological tests and they can be qualified as highly standardised since the same standard materials are used at each application.

To date about a dozen different Toxkits have been completed and commercialised. An overview of the Toxkit microbiotests which are used in freshwater and terrestrial ecotoxicology is given in the Table in annex. A battery of Toxkits is now thus available with micro-algae, ciliated protozoans, rotifers and several crustaceans for performance of acute and/or short chronic assays, with exposure times ranging from 24h up to 7 days.

Several Toxkit microbiotests listed in the Table make use of the test species prescribed by the regulatory organizations for conventional bioassays, e.g. the micro-algae *Raphidocelis subcapitata* and the crustaceans *Daphnia magna* and *Daphnia pulex*, and *Ceriodaphnia dubia*. During the development of these specific Toxkit assays efforts have been made to adhere to the experimental procedures prescribed by the OECD and the ISO. The only difference with the corresponding conventional bioassays is that the Toxkits contain the dormant or immobilised test biota, which can be hatched or set free « on demand » at the time of performance of the assays, thus bypassing all the problems and costs associated with stock culturing.

As demonstrated in recent papers (Persoone, 1998a and 1998b; Fochtman 2000; Latif and Zach 2000; Ulm *et al.* 2000; Vandenbroele *et al.* 2000; Van der Wielen and Halleux 2000, Galassi and Croce, 2000) the sensitivity of the Toxkits for pure chemicals as well as for multi-chemical products or natural samples is very similar to that of the corresponding conventional bioassays.

4   TOXKIT MICROBIOTESTS FOR ROUTINE TOXICITY TESTING AND MONITORING

The development of low cost microbiotests with biota from different trophic levels and their commercial availability, has rapidly triggered their incorporation in research and in toxicity monitoring programmes in several countries. Toxkit assays are now used for investigations and routine applications in a variety of domains encompassing pure chemicals, industrial products and effluents, wastes and waste leachates, surface waters and groundwaters, contaminated soils and their run-offs and percolates, and even biotoxins. The state of the art on the development and application of microbiotests was evaluated two years ago in the International Symposium on « New Microbiotests for Routine Toxicity Screening and Biomonitoring » during which more than 100 oral and poster papers were presented, 60 of which were published in the Symposium Proceedings (Persoone *et al.* 2000). As of to date nearly 100.000 Toxkit microbiotests have already been performed in governmental, industrial and research laboratories in 40 countries worldwide and more than 100 papers with Toxkit data have been published in scientific literature.

5 APPLICATION OF TOXKIT MICROBIOTESTS FOR TOXICITY EVALUATION OF OILS AND LUBRICANTS

Like for any type of multi-chemical samples of which the real hazard can only be determined with biological techniques, Toxkit microbiotests can also be applied for the determination of the toxicity of oils and lubricants.

As is the case for most products with a low water solubility, the hazard of oils and lubricants products will nevertheless be coming in the first place from the chemicals which it contains, which are "bioavailable", i.e. those compounds which can "leach out" (dissolve) in water.

OVERVIEW OF THE TOXKIT MICROBIOTESTS AVAILABLE TO DATE WITH DIFFERENT FRESHWATER TEST SPECIES

Toxkit	Type of biota	Test species	Inert stocks	Type of test	Test duration	Test criterion
Algaltoxkit F™	Microalgae	*Raphidocelis subcapitata*	algal beads	chronic	72 h	growth inhibition
Protoxkit F™	Ciliated protozoa	*Tetrahymena thermophila*	stationary cultures	chronic	24 h	growth inhibition
Rotoxkit F™ acute	Rotifers	*Brachionus calyciflorus*	cysts	acute	24 h	mortality
Rotoxkit F™ short chronic	Rotifers	*Brachionus calyciflorus*	cysts	chronic	48 h	reproduction
Thamnotoxkit F™	Anostracan crustaceans	*Thamnocephalus platyurus*	cysts	acute	24 h	mortality
Daphtoxkit F™ magna	Cladoceran crustaceans	*Daphnia magna*	ephippia	acute	48 h	immobility/mortality
Daphtoxkit F™ pulex	Cladoceran crustaceans	*Daphnia pulex*	ephippia	acute	48 h	immobility/mortality
Ceriodaphtoxkit F™ acute	Cladoceran crustaceans	*Ceriodaphnia dubia*	ephippia	acute	24 h	mortality
Ostracodtoxkit F™ chronic	Ostracod crustaceans	*Heterocypris incongruens*	cysts	chronic	6 d	mortality and growth inhibition

A technology which is currently applied for oils and lubricants consists in mixing a certain volume of the product with a certain volume of water, followed by the determination of the toxicity of the water fraction to specific test species.

Such a leaching out procedure has been applied successfully over the last two years by BFB Oil Research in Namèche in Belgium, to whom we refer the reader for more detailed information.

6 CONCLUSIONS

Taking into account that, besides the advantages outlined above, performance of toxicity tests with Toxkits decreases the costs by more than one order of magnitude in comparison to conventional bioassays, the commercial availability of a battery of acute and short-chronic culture/maintenance free and user friendly microbiotests has now opened the door for routine ecotoxicological testing in many fields. Thanks to these cost-effective microbiotests, determination in routine and biomonitoring of the hazards of multi-chemical industrial products including oils and lubricants and their wastes, is now also possible.

Hopefully this new tool will be gradually exploited to its full potential for this specific category of industrial products, not only for the protection of environmental biota, but also for man who, as said above, is ultimately totally dependent of the well-being and the normal functioning of the ecosystems in which he lives daily.

References

1. Blaise, C. (1991). Microbiotests in aquatic ecotoxicology : characteristics, utility and prospects. *Environmental Toxicology and Water Quality* : an international Journal. 6, 145-155.
2. Bulich, A.A. (1979). Use of luminescent bacteria for determining toxicity in aquatic environments. *In : Aquatic Toxicology* : Second Conference. L.L. Marking and R.A. Kimerle (Eds). ASTM STP 667. American Society for Testing and Materials, Philadelphia PA., 98-106.
3. Fochtman, P. (2000) Acute toxicity of nine pesticides as determined with conventional assays and alternative microbiotests. *In : New Microbiotests for Routine Toxicity Screening and Biomonitoring*, Persoone, G., Janssen, C., and De Coen, W. (Eds). Kluwer Academic/Plenum Publishers, 233-241.
4. Galassi S. and Croce V., (2000) Test acuto con Daphtoxkit F magna per la valutazione della tossicità di un effluente industriale e l'individuazione dei composti tossici. Biologia Ambientale 14 (2), 21-28.
5. Latif, M., and Zach, A. (2000). Toxicity studies of treated residual wastes in Austria using different types of conventional assays and cost-effective microbiotests. *In : New Microbiotests for Routine Toxicity Screening and Biomonitoring*, Persoone, G., Janssen, C., and De Coen, W. (Eds). Kluwer Academic/Plenum Publishers, 367-383.
6. OECD (1984). Terrestrial plants, growth test. OECD Guidelines for the Testing of Chemicals. N° 208. *Organization for Economic Co-operation and Development*, Paris.
7. Persoone, G., and Van de Vel, A. (1987). Cost-analysis of 5 current aquatic ecotoxicological tests. *Report EUR 1134 EN*. Commission of the European Communities.
8. Persoone, G. (1991). Cyst-based toxicity tests. I. A promising new tool for rapid and cost-effective toxicity screening of chemicals and effluents. *Zeitschr. für Angew. Zoologie* 78, 235-241.

9. Persoone, G. (1998a). Development and first validation of a "stock-culture free" algal microbiotest : the Algaltoxkit. *In : Microscale Testing in Aquatic Toxicology. Advances, Techniques and Practice*, Wells, P.G., Lee, K., and Blaise, C. (Eds), C.R.C. Publishers. Chapter 20, 311-320.

10. Persoone, G. (1998b). Development and first validation of Toxkit microbiotests with invertebrates, in particular crustaceans. *In : Microscale Testing in Aquatic Toxicology. Advances, Techniques and Practice*, Wells, P.G., Lee, K., and Blaise, C. (Eds), C.R.C. Publishers. Chapter 30, 437-449.

11. Persoone, G., Janssen, C., and De Coen, W. (Eds). 2000. *New Microbiotests for Routine Toxicity Screening and Biomonitoring* Kluwer Academic/Plenum Publishers, 565 pages.

12. Ulm, L., Vrzina, J., Schiesl, V., Puntaric, D., and Smit, Z. 2000 Sensitivity comparison of the conventional acute *Daphnia magna* immobilization test with the Daphtoxkit F^{TM} microbiotest for household products. *In : New Microbiotests for Routine Toxicity Screening and Biomonitoring*, Persoone, G., Janssen, C., and De Coen, W. (Eds). Kluwer Academic/Plenum Publishers, 247-252.

13. Vandenbroele, M.C., Heijerick, D.G., Vangheluwe, M.L., and Janssen, C.R. 2000. Comparison of the conventional algal assay and the Algaltoxkit F^{TM} microbiotest for toxicity evaluation of sediment pore waters. *In : New Microbiotests for Routine Toxicity Screening and Biomonitoring*, Persoone, G., Janssen, C., and De Coen W. (Eds). Kluwer Academic/Plenum Publishers, 261-268.

14. Van der Wielen, C. and Halleux, I. (2000). Shifting from the conventional ISO 8692 algal growth inhibition test to the Algaltoxkit F^{TM} microbiotest ? *In : New Microbiotests for Routine Toxicity Screening and Biomonitoring*, Persoone, G., Janssen, C., and De Coen, W. (Eds). Kluwer Academic/Plenum Publishers, 269-272.

15. Wells, P.G., Lee, K., and Blaise, C. (Eds) (1998). Microscale Testing in Aquatic Toxicology. Advances, Techniques and Practice. CRC Publishers. 679 pages.

USE OF *LUX* BACTERIAL BIOSENSORS FOR RAPID TOXICITY DETECTION AND PROTECTION OF SEWAGE TREATMENT PROCESSES

K. Killham, A. M. Horsburgh, D. P. Mardlin, I. Caffoor*, L. A. Glover**, and M. S. Cresser.

Departments of Plant and Soil Science and **Molecular and Cell Biology
University of Aberdeen, Aberdeen AB24 3UU
*Yorkshire Water Services
Halifax Road, Bradford BD6 2LZ

1 INTRODUCTION

In the UK, the wastewater treatment industry requires a rapid assay for effluent water quality. Toxic shocks to sewage treatment works represent an important economic cost to the water industry and can cause adverse environmental impacts. Rapid and reliable on-line testing is required to protect biological treatment and thus to reduce the industry's current overspend (approx. 10%) because of industrially derived toxic effluents restricting the microbial efficiency of biological sewage treatments works.[1] Luminescence-based microbial biosensors appear to offer a very promising way forward for assessing the toxicity of water and wastes.[2,3] The development of *lux*-based bacterial biosensors enables the sensitive and ecologically relevant detection of pollutants to be linked to luminometry for batch and on-line toxicity testing instrumentation (NEWT – novel ecotoxicity water test). This paper describes the use of an on-line monitor using the luminescent response of bacterial biosensors as an indicator of toxicity to sewage works.

2 MATERIALS AND METHODS

2.1 Biosensor Construction and Use

Biosensors were constructed through the cloning of *lux* genes into terrestrial host bacteria. The organism *E. coli* HB101 was *lux*-marked with the plasmid pUCD607 carrying the full *lux* CDABE cassette. In this construct, all the genes necessary for light production are present, and light is emitted during the following reaction (Figure 1).

Figure 1 *Mechanism of light production by the biosensor.*

Prior to toxicity testing, the biosensors were freeze-dried and stored at -20°C. The biosensor toxicity test involved the exposure of resuscitated freeze-dried cells to the test aqueous sample and measuring the light response after selected exposure times varying from 5 to 15 minutes.

2.2 Sample Preparation and Toxicity Testing

Fresh, activated sewage samples were collected from a sewage treatment works with minimal industrial inputs. The samples were diluted with deionised water and selectively spiked with 3,5 dichlorophenol (DCP) in concentrations ranging from 0 to 100 mg l^{-1} or Zn in concentrations ranging from 0 to 50 mg l^{-1}. The samples were analysed using both an on-line toxicity test and electrolytic respirometry. The on-line instrument was developed specifically to accommodate *lux* bacterial biosensors and to monitor the light output response to samples at 6 second intervals throughout the luminometric measurement. The system consisted of a pump, a dilution system and a unit for light detection connected to a PC. The biological oxygen demand (BOD) of the same samples was analysed using a Merit 20 electrolytic respirometer (A. E. Instruments).

2.3 Use of Model Sewage Treatment System

A model sewage treatment system was set up and supplied with raw sewage (Figure 2).

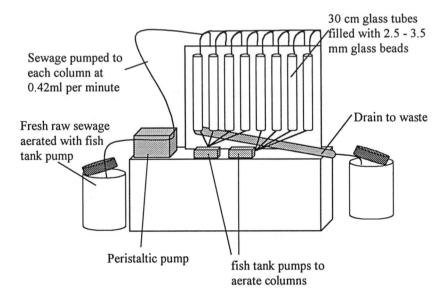

Figure 2 *Model Sewage Treatment System.*

After a period of equilibration (3 weeks), the columns were supplied with sewage spiked with a range of concentrations of the standard toxicant 2-bromo-2-nitro-1,3-propanediol (HOCH$_2$C(Br)(NO$_2$)CH$_2$OH) (Bronopol) for 2 hours. Unspiked sewage was supplied to the system for a further 2 hours. Samples of sewage were taken before spiking with Bronopol, and after, and transferred to the electrolytic respirometer for BOD analysis and the on-line toxicity monitor for toxicity testing.

3 RESULTS

3.1 Toxicity Testing

Both the on-line monitor and the respirometer detected toxicity in the sewage spiked with Zn and 3,5 DCP. The bioluminescence response to toxicity showed a strong linear relationship with the respiration response of Zn spiked sewage (Figure 3).

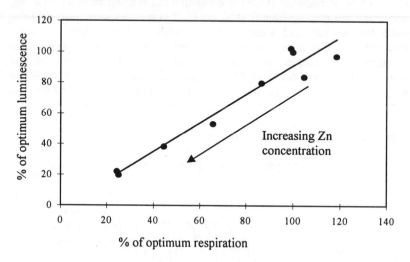

Figure 3 *Respiration response versus luminescence response – Zn spiked sewage.*

The corresponding relationship between luminescence and respiration response for 3,5 DCP spiked sewage was more complex, but equally reproducible (Figure 4).

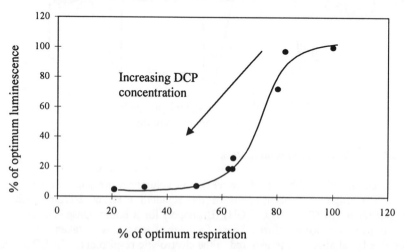

Figure 4 *Respiration response versus luminescence response – 3,5 DCP spiked sewage.*

3.2 Model Sewage Treatment System

After the 3 week equilibration period, the model sewage treatment system significantly reduced the BOD of the sewage. There was an average reduction of 75 to 85% in the BOD after treatment. When the system was supplied with sewage spiked with high concentrations of Bronopol, the BOD after sewage treatment was higher, indicating that there had been a reduction in the ability of the system to treat sewage (Table 1). The effect was significant in the columns spiked with 10 mg l^{-1} Bronopol.

Table 1 *BOD of sewage following toxic insult of sewage treatment system.*

Concentration of Bronopol in sewage supplied to treatment system	BOD expressed as % BOD of untreated sewage
No sewage treatment	100
0 mg l^{-1}	15.0
0.75 mg l^{-1}	13.6
2.0 mg l^{-1}	14.7
5.0 mg l^{-1}	20.9
10.0 mg l^{-1}	29.6

The on-line toxicity monitor reliably detected the toxicity of Bronopol in sewage (Figure 5). The monitor detected an EC_{50} value for Bronopol in sewage of 3 – 3.5 mg l^{-1}. This is a concentration relevant to the pollutant loadings experienced by the water industry.[1]

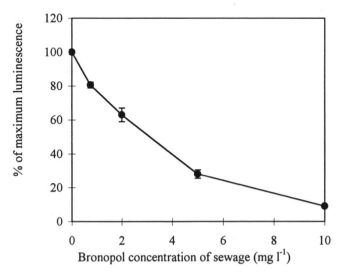

Figure 5 *Toxicity of sewage spiked with Bronopol as measured using the on-line toxicity monitor (Mean values +/- SE of mean).*

4 CONCLUSIONS

The biosensor was shown to be a good predictive indicator of pollutants toxic to the sewage treatment process. A range of environmentally relevant biosensors is now available, and includes *lux*-marked pseudomonads isolated from sludge, soil and water.[3] Electrolytic respirometry has previously been used for toxicity testing and comparison with luminescence based toxicity testing.[4,5] In this study, the results from the on-line toxicity monitor correlated well with respirometry results, for both the heavy metal and the organic pollutants tested. The biosensor test was as sensitive to the presence of Zn, and more sensitive to the presence of DCP in the sewage when compared to the respirometric test. The differences in the relationships between luminescence and respiration may indicate different modes of toxic action for heavy metal and organic pollutants. The on-line instrument showed little biofouling after prolonged testing of raw, unfiltered sewage and the toxicities of the toxicants tested remained constant over several hours.

The use of the model sewage treatment system demonstrated that the on-line toxicity monitor reliably detected levels of the toxicant Bronopol that caused a significant reduction in the efficiency and effectiveness of the biological treatment process. The bacterial biosensor test was sensitive enough to detect levels of pollution well below the level shown to cause a detrimental effect to the treatment process, indicating that the test offers an excellent early warning device for the protection of biological treatment processes.

The *lux* biosensor proved to be a reliable predictive test for toxic inhibition of biological treatment for both organic and heavy metal pollutants. This indicates that NEWT offers the water industry a powerful tool for the efficient protection of biological treatment processes as well as offering an environmentally relevant toxicity test, with the rapidity and sensitivity required to serve as a benchmark direct toxicity assessment for a wide range of waste and freshwater applications.

Acknowledgements

The authors wish to acknowledge the help of the following: BBSRC, Yorkshire Water Services, Siemens Environmental Systems, BPB Davidsons Paper and J&G Grant.

References

1. I. Caffoor. Pers. comm. 1995.
2. G. I. Paton, C. D. Campbell, L. A. Glover and K. Killham. Letters in Applied Microbiology, 1995, **20**, 52-56.
3. G. I. Paton, E. A. S. Rattray, C. D. Campbell, M. S. Cresser, L. A. Glover, J. C. L. Meeussen & K. Killham. 'Biological Indicators of Soil Health' Eds. C. F. Parkhurst, B. M. Doube and V. V. S. R. Gupta. CAB International, Oxford, 397-418, 1996.
4. J. S. Brown, E. A. S. Rattray, G. I. Paton, G. Reid, I. Caffoor and K. Killham. Chemosphere, 1996, **32**, 1553-1561.
5. E. A. S. Rattray, D. Fearnside, I. Caffoor and K. Killham. Environmental Technology, 1996, **6**, 8.

RAPID DETECTION OF RESIDUAL CLEANING AGENTS AND DISINFECTANTS IN FOOD FACTORIES

Juha Lappalainen[1,4], Satu Loikkanen[2], Marika Havana[3], Matti Karp[1], Anna-Maija Sjöberg[2] and Gun Wirtanen[2]

[1]University of Turku, Department of Biotechnology, [2]VTT Biotechnology and Food Research, Espoo, Finland, [3]University of Turku, Department of Biochemistry, [4]Bio-Nobile Oy, Turku, Finland

1 INTRODUCTION

ATP bioluminescence measurement has been widely used for the detection of microbial contamination and food residues in the food industry. ATP measurements and culture methods very often correlate well with cultured laboratory samples, but the situation is less favorable, if actual samples from the industry are used.

Disinfectants can have severe effects on ATP measurement and especially on the enzyme, firefly luciferase, which is used in the commercial tests.[1] The enzyme and measurement itself are readily affected by disinfectants, which can have a quenching or even additive effect on the measurement. The noneffective concentration of the agents on the ATP measurement is usually well below the ready-to-use concentration of the disinfectants used.

An ATP standard is always used to quantify the total ATP in starter cultures, but in hygiene testing the market requires as simple method as possible, and this has led to a situation where no internal standard is used. This makes the methods easy to perform, but it can also have negative effects on the reliability of the test result. Different chemicals show very different behavior on ATP measurement and therefore the results can be totally unpredictable.

The chemical residues of 38 cleaning agents and disinfectants of various types were measured using two microbiological methods. The methods were the inhibition zone technique and the luminescence based-toxicity method.[2] The luminescence method is widely used in environmental toxicity testing and is based on the standardized *Vibrio fischeri* test.[3] The aim of this study was to show that small amounts of disinfectants and cleaning agents could be detected with this microbial residue testing method. A small case study was performed to estimate the presence of the chemicals before starting the production.

2 MATERIALS AND METHODS

2.1 BioTox test

The luminescence test was carried out using the *V. fischeri* bacterium from the BioTox kit (Bio-Orbit, Turku, Finland) and a 1253 luminometer or Sirius luminometer (Bio-Orbit). The bacteria produce light as part of their natural metabolism. Toxic substances, e.g. tensides, interfere with these metabolic processes, resulting in a reduction of the light output. The protocol was a standard DIN 38412 protocol, with the exception that the pH was not adjusted for the *V. fischeri* test. An overview of the method is described below.

The light-producing bacteria *V. fischeri* were reconstituted from the freeze-dried vials and stabilized for 1 hour at +4 °C. A 500 µl aliquot of *V. fischeri* suspension was transferred into the measuring cuvette, and the light output was measured with a luminometer. This was performed for both the control tube and for the sample tubes. After this first measurement, 500 µl sample of detergent dilutions were mixed together with the bacteria and the tubes were incubated at +15 °C for 5 min. The second reading was taken after the incubation, and the light production was measured from the control and samples. The inhibition in light production was calculated and compared to the light output of unstressed control bacteria. The results are expressed as inhibition percentage (INH%), which was automatically calculated with BioTox software.

The test was performed from liquid samples and from dried surface samples. The dried surface samples were made my spreading 1 ml of the chemicals over a stainless steel plate. The plate was then dried overnight in room temperature and before the measurement 1.5 ml of distilled water was added to the surface and the surface gently rubbed with a pipette tip. A plate with distilled water was used as a control.

2.2 Inhibition zone technique

The inhibition zone technique was performed using filter paper disks of (Schleicher & Schüll 2668 Antibiotica Testplättchen) soaked in various dilutions of chemicals. Three disks were placed on top of each other on tryptone soy agar plates inoculated with *Micrococcus luteus* VTT E-91474. The plates were incubated at 37 °C for 18 h. The inhibition zone formed was measured with a ruler. A total of 38 commercial cleaning agents and disinfectants were tested using different dilutions, and the results were compared to the control sample containing no disinfectants. All tests were run in triplicate.

2.3 Study in a food factory

A small study was performed to determine if detergents or disinfectants were present on food factory equipment surfaces after rinsing. The samples were taken after normal washing immediately before production began in the facilities. The sample was taken with a pipette tip from the droplets on the surface. The sample volume was 500 µl. A total of 31 randomly selected samples were taken, and the inhibition was measured with the BioTox test. The reference sample with this test was the tap water that was used for the washing and rinsing in the factory.

3 RESULTS

3.1 Inhibition zone technique and BioTox test

Results from the different types of chemicals tested with the inhibition zone and bacterial luminescence tests are presented in Table 1. The BioTox method appears to be extremely

Table 1 *Example of the luminescence and the inhibition zone method results with different chemicals. The BioTox (luminescence) method showed very good sensitivity in the residue testing and it was adaptable also for dry surfaces.*

Chemical	Effective agent	Ready-to-use concentration (%)	Liquid samples Inhibition Zone	BioTox	Surface samples BioTox
			Visible inhibition in concentration (%)	Concentration giving 50% inhibition (%)	Percentage of inhibition with ready-to-use concentration
1	Citric acid, persulphate	1	1	0.002	99.9
2	Nitric acid	0.5	2	0.02	95
3	Ethanol	100	ND	5.9	59.8
4	Isopropanol	100	>1	0.5	99.9
5	Organic alcohol, soap	0.5	ND	0.01	94.2
6	Hydrogen peroxide, peracetic acid	0.3	0.5	3×10^{-6}	1.3
7	Sodium hypochlorite	0.2	0.5	0.0003	40.7
8	Sodium hydroxide	2	2	0.008	99.9
9	Quats	2	0.2	0.008	99.9
10	Sodium hydroxide	3	1	0.004	99.9

sensitive to all types of cleaning agents and disinfectants but the inhibition zone technique is much less sensitive. The inhibition zone technique was not sensitive enough to be used with the dried surface samples, and there are only results for the liquid samples in Table 1. With the BioTox test it was shown, that chemical number 6 evaporated from the surface sample, while all other chemicals left residues on the surface.

When the BioTox results were compared with those presented in the literature, pure chemicals expressed effects very similar to those of the cleaning agents used here.[4] It must be noted that the manufacturers do not provide information on all chemicals and their precise concentrations; instead they are usually expressed as ranging from 5% to 15%, for example. Cleaning agents are often complicated mixtures of chemicals, and even small amounts of very toxic substances cause a pronounced change in the toxicity estimation.

3.2 Study in a food factory

The results in a food factory are presented in Figure 1 and results show that there is a risk of chemical contamination caused by the disinfecting agents from washed surfaces which are in

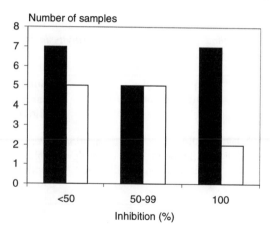

Figure 1 *Study of the disinfectant residues in a food factory. Samples were divided into three groups according to the inhibition obtained with modified* Vibrio fischeri *test. Total number of samples 31. Direct contact with the food (■), no contact with the food (□).*

direct contact with the food. The samples belonging to the highest inhibition class probably contained high concentrations of chemicals, which was really undesirable in this case. This also suggests that severe problems may arise in the facilities if the ATP testing is performed without any control system to ensure the proper action of the reagents and to detect possible remaining disinfectant residues on the surfaces.

4 DISCUSSION

The luminescence bacteria light inhibition method (*Vibrio fischeri* test) can be used to measure low amounts of residues both in liquids and on surfaces. This method is rapid and easy to perform. The detection limit for the inhibition zone technique (*Micrococcus luteus* assay) was often as high as or near the ready-to-use concentration. Therefore this method is not sensitive enough to eliminate the possible error in the ATP measurement caused by the disinfectants. Further, on the contrary to the luminescent method which gives and answer during food processing, this cultivation method is not applicable to actual real life situation in process lines.

Chemical residues allowed in processed food usually occur at very low level, but no testing is actually done before starting production. The results of this study show that there is a real possibility of residues remaining. Even the disinfectants, that are according to the instructions evaporated from the surface, leave residues that can be detected with the luminescence method. This type of testing is also important because of the possibility of resistant bacteria, which is already a problem in hospitals.

The luminescence method can be applied to liquids, and also to dried surfaces. The method is unspecific to the disinfectants and detergents used, and therefore it can be used for screening purposes, when the total cleanliness of the food processing is evaluated.

Acknowledgements

The financial support from Maj and Tor Nessling foundation is gratefully acknowledged.

References

1. T.A. Green, S.M. Russell and D.L. Fletcher: Effect of chemical sanitising agents on ATP bioluminescence measurements. *J. Food Prot.,* 1998, **61(8)**:1013-1017.
2. A.A. Bulich and D.L. Isenberg: Use of luminescent bacterial system for the rapid assessment of aquatic toxicity. *ISA transactions*, 1981, **20**, 29-33.
3. ISO/CD (draft) 11438 (1994) Water quality. Determination of the Inhibitory Effect of Water Samples on the Light Emission of *Vibrio fischeri* (Luminescent bacteria test)
4. K.L.E. Kaiser and J. Devillers: Ecotoxicity of Chemicals to Photobacterium phosphoreum 1994. Handbooks of Ecotoxicological Data, Gordon and Breach Science Publishers.

Rapid Methods in Food Microbiology
Dr. Stuart. A. Clark

The food and food ingredients microbiology testing market is very large, greater than £ 2.0 billion in Europe alone. The vast majority of this testing is still performed using traditional microbiology methods involving the culture of bacteria in culture broths and on agar plates. The bacterial colonies are than identified by morphological characteristics, biochemical utilisation profiles or immuno-detection methods such as latex agglutination. These traditional microbial culture methods have been developed over many years and been shown to be very sensitive, selective and reproducible in many laboratories across the globe.

The question to be addressed by proponents of rapid methods must be why do we need rapid methods and by what criteria do we determine if they present a significant improvement on the established culture methods?

The three most common attributes of rapid methods used to promote their use are :

- Time savings
- Cost savings
- Increased testing capacity

By definition rapid methods should deliver results faster than conventional culture methods (or they are not considered rapid). If a laboratory can cut the amount of labour work per test this will obviously decrease their cost per test. Finally if the rapid method can be performed in larger numbers than traditional culture methods, perhaps even be automated, then the laboratory can greatly increase its testing capacity thereby enabling it to take on more work with the same numbers of staff.

These various arguments are often appealing to the testing laboratories, however, from their point of view there are a number of other factors which influence their decisions in terms of switching over to a new rapid method :

- Regulatory Approval
- Staff Training
- Method Validation in Laboratory
- Operation of New Method in Current Laboratory Work Schedules

Rapid Methods must offer a confirmed result equivalent to the current traditional culture method. This needs to be demonstrated by an appropriate regulatory approval trial, which can be cited by the laboratory when they are audited. To ensure the rapid method delivers the expected results in the laboratory it is vital that the staff involved are properly trained and can demonstrate that they are competent to perform the new method. Before the laboratory can start using the new method they need to validate that it performs as expected in their laboratory. Finally, the successful integration of the new method will only occur if it can be fitted into the routine schedules of the laboratory.

With all of these hurdles to overcome, the suppliers of new rapid methods have to convince the laboratories that the savings to be made are worth all of the effort required to establish the new rapid methods in their laboratory. This is starting to happen but it is a slow process. It may, therefore, be a many years before rapid methods become the norm rather than the exception in food testing laboratories.

FOOD SAMPLE PREPARATION AND ENRICHMENT FOR RAPID DETECTION

A. N. Sharpe

PO Box 1224
Almonte
Ontario K0A 1A0
Canada

1 INTRODUCTION

A challenge to make analytical chemists gape is the requirement by agencies that regulate the microbiological quality of food, to detect, say, one *Salmonella* cell in 25g of food. It represents a limit of detection of $1:10^{12\text{-}13}$ for the cell itself, and only $1:10^{18\text{-}19}$ for, say, a nucleotide sequence characterizing the genus. Detecting chemical analytes at this level requires a MS-GC, or similar expensive apparatus. Food microbiologists manage it daily, using a few ££ of materials, letting microbial cells *multiply* (amplify), essentially noise-free, by factors of $\geq 10^7$. Colonies (agar) or broth suspensions resulting from incubation consist of millions of cells, signals detectable by eye or by a variety of manual or automated chemical, biochemical, immunological or DNA-based tests. Techniques for enriching pathogens by *incubation* are very selective, and have the required limit of detection; their disadvantage is the 24-96h incubation incurred. Nowadays, perishable foods are transported around the world, cold-storage warehousing costs may require rapid turnover of stocks, etc. The safety of foods must be ascertained more rapidly. To meet demands for shorter analytical times, scientists are developing ways to avoid the lengthy incubations that yield large microbial populations, by substituting physical or chemical enrichment for microbial growth.

2 THE ANALYTICAL PROBLEM OF RAPID MICROBIAL DETECTION

A food microbiologist faced with confirming the *absence* of a microbial pathogen in 25g of food by traditional methods can incubate the whole sample in broth in the assurance that, if there is at least one target cell, even one firmly attached to the sample tissue (biofilm), it will multiply and spread into the broth during incubation and eventually be detectable.

Detecting or quantifying foodborne microbes *without* incubation is a bigger challenge. To detect, one must get microbes from a food into the detection system. Ideally, one might detect them *in* the food by conferring on them light emission, radioactivity, or other property detectable against the background of the food. Currently one must extract whole cells and detect them by serological, enzyme, or other phenotypic property; or extract components (DNA or RNA) to provide a consistent analytical base. Even today there are no sampling methods capable of removing microbes quantitatively from foods. For physical or chemical separation and concentration one has little alternative to first suspending the food in 100-250 ml of diluent, in order to have a modest probability of liberating the target cell for detection.

Microscopy, flow cytometry and other *direct* techniques currently are useless at regulatory limits of detection. At the high magnification needed to observe bacterial cells one needs 10^{6-7} cells/ml to have any practical probability of getting a target cell into a microscope field. With just one target cell in 250 ml, the limit of detection is far away. The situation is very little better for immunological, DNA-based or other analyses.

The detection problem is not actually in achieving microbial populations of $\geq 10^6$ cells, but of $\geq 10^6/ml$ (i.e., the essence is *concentration* not *number*). Tests *can* be made on microscopic volumes *if* target cells can be persuaded to be in them (i.e., can be shifted into smaller volumes). However, if an initial suspension of 250 ml contains only a single target cell (regulatory limit) the whole volume must be "enriched" physically or chemically in order to detect it. Just *how* distant the limit of detection is, can be seen by considering a single high-power microscope field (with the *potential* for identifying microbial cells on the basis of a phenotypic property). With a (focused) volume of around 10^{-9} ml, to ensure that target cells from 250 ml of suspension arrive in a field of view, one must concentrate the sample through a factor of $10^{11}:1$. Microscopy is an extreme example, but serves as a good illustration of the problem.

A miscellany of separation/concentration techniques have been investigated, among them: centrifugation; membrane filtration; immunomagnetic particles; ion-exchange; affinity chromatography; biphasic partitioning; electro- or dielectrophoresis; standing-wave ultrasound; and foam flotation. Some are useful in limited situations, but few are practical at volumes greater than 1-2 ml, through cost (e.g., antibodies), short range (magnetic or electric fields), or other limitations. At present, membrane filtration (Direct Epifluorescent Filter Technique, or DEFT) and immunomagnetic particles are the most useful.

Currently, the most successful techniques combine a 6-24 h broth incubation, during which the target multiplies, capture on a selective substrate (e.g., immunomagnetic beads), then detection by, say, polymerase chain reaction (PCR). Compromise methods like these are used in several commercial "ultra-rapid" test kits. I mainly discuss routes to detecting microbial pathogens directly, without recourse to multiplication, although the *ultimate* goal of instantly detecting microbial pathogens in foods may lie far in the future.

Two major avenues urgently need study: i) methods to cheaply collapse primary sample volumes into a few ml so that techniques capable of more specific separations, but currently too costly for large-scale use, can be applied, and; ii) techniques to produce food sample suspensions that are inherently better suited to separating out the target cells.

3. SEPARATION AND CONCENTRATION TECHNIQUES

3.1 Multiplexed Separations

Before discussing individual techniques, I want to emphasize how practical procedures to achieve large concentration factors might combine a multiplicity of less efficient steps. A simple model suggests that multiplexed steps could yield separations dramatically *faster* than single-step processes, at the same time attacking the *volume/cost* problem. The philosophy should be kept in mind by anyone developing rapid detection procedures.

The argument[1] is general; one could consider any separation process (centrifugation, membrane filtration, electrophoresis, ion-exchange, affinity chromatography, flotation, etc) and any detection method (microscopy, flow-cytometry, CCD/luminescence, etc). Without worrying about which techniques to use for the separation stages, how long does it take us if we try to extract target cells in a single stage, or two or more less ambitious stages?

Single-Stage Process: Consider a single-stage concentration process as the operation of passing a *capture element* (in reality, the final volume) through a sample N times greater in volume, until it has passed through the whole sample and captured the target microbes from it. If it takes t seconds for the capture element to pass through its own volume in the sample (*specific sweep time*) and this remains constant during the capture pass, the time T required for the capture element to reach the end of its pass is:

$$T = t(N - 1) \qquad\qquad 1$$

Two-Stage Process: Now concentrate the sample in two stages, each of which is less efficient (yields a smaller increase in analyte concentration) than the above. Instead of just one, use n capture elements for the first stage and, after concentrating the sample into these, concentrate these n elements into one in the manner above. The total time for concentration is:

$$T = (T_1 + T_2) = t(n + \frac{N}{n} - 2) \qquad\qquad 2$$

and is minimum when $n = N^{1/2}$, i.e., when the overall concentration factor is achieved in two stages of approximately equal effectiveness.

Three or More-Stage Process: A three-stage process is a two-stage process with a stage added at the front-end. If t is the same for all three processes the total time needed is:

$$T = t(2n + \frac{N}{n^2} - 3) \qquad\qquad 3$$

and is minimum when $n = N^{1/3}$, i.e., three stages of approximately equal effectiveness yield the minimum concentration time. Similarly, p multiplexed processes execute in a minimum time when $n = N^{1/p}$.

Relative Separation Times: To give perspective to this imagine an extreme case of concentrating pathogens from 1,000 ml of suspension into one high-power microscope field (10^{-9} ml), ignoring practicalities such as the need to prepare the product of one stage for introduction to the next (e.g., eluting from a column). To derive a plausible value for t one might consider membrane filtration, where a 1 ml sample might be reduced to 1.10^{-4} ml when it is captured as a layer approximately 1 μm thick in 10s, giving t = 1/1000s for this process. Other processes (e.g., antibody-coated beads) have different specific sweep times, but we can ignore it in illuminating the relative efficiency of multiplexed procedures. By using t = .001s for all stages in concentrating the microbes from a 1,000 ml sample into a 10^{-9} ml microscope field, we find that for single-, two- and three-stage processes, total concentration times would be 1.10^9s (32 years), 2.10^3s (33 min), and 30s, respectively.

Even allowing for different specific sweep times of different processes, the effect of multiplexing separation stages, shown by this simple treatment, *is so dramatic* that one should keep it in mind. The possibilities are endless. One might, for example, combine a large column as first stage (plus washing and elution), second stage centrifugation, a third stage in the flow-cytometer, and so on. Or, a process might be repeated on a smaller scale. As sample volume reduces through the stages, adsorption to antibody-coated filaments or pin-point areas on microscope slides, might be considered. *The important thing is not to expect too much of any one stage or the overall process slows down.* I emphasize that this is not a solution, just a philosophical guide to process development.

3.2   Membrane filtration separations

This prime separation method is capable (not necessarily without difficulty) of separating microbial cells from food tissues and concentrating them for detection. It can lower limits of detection in direct microscopic methods by several orders of magnitude, for example in the Direct Epifluorescent Filter Technique[2]. In a related procedure *E. coli* O157 has been detected directly in juice and meat at 10 cells/g^3.

The utility of filtration depends very strongly on the filterability of food suspensions. Suspensions of unprocessed foods (raw meats, fish, vegetables) generally filter easily. With increasing levels of processing, gums, fillers, etc, food suspensions become less filterable, and dairy products are often a problem. A great deal of work on improving the membrane filterability of food suspensions without killing bacteria was carried out during development of the hydrophobic grid membrane filter[4-6]. For microscopy, where the ability of cells to multiply is not important, digestion with trypsin and Triton X-100 surfactant efficiently removes unwanted debris[2,7,8].

3.3   Centrifugation

Despite being inconvenient, centrifugation will probably continue to have a major role in microbial separation. Centrifugation at 2,000 g for 10 s prior to estimating biomass by ATP measurement removed virtually all meat particles without loss of bacterial count[9]. Density gradient centrifugation can remove food debris without loss of bacterial count; a 15 min method using colloidal silica clarified food suspensions[10], and an automated density gradient apparatus in the Bactoscan instrument also allows concentration of food-related microbes[11]. Sedimentation Field-Flow Fractionation separates pure bacterial cultures and is probably applicable to foods; cells are injected into an open, unpacked channel, first sedimented by a low (5-10 RCF) centrifugal field, then fractionated by a parabolic fluid-velocity field as diluent passes through the chamber[12].

3.4   Biphasic partitioning

The tendency of bacteria, viruses, and other bodies to partition themselves between the phases of aqueous *biphasic* systems (e.g., of polysaccharide and gelatin mixtures), permits some degree of separation[13]. Not only can *Salmonella* and *Escherichia coli* be separated, but also rough/smooth mutants *of Salmonella typhimurium*[14].

3.5   Dielectrophoresis

Conducting particles, suspended in liquid in a non-uniform electric field between a plate and a pin electrode, migrate to or from the pin electrode. The direction of movement depends on the relative conductivities of particle and liquid. Unlike *electrophoresis*, which needs a DC field, dielectrophoresis occurs in AC or DC fields. Electrode assemblies are barely larger than microbial cells, and may be fabricated on silicon semiconductor chips. Complex electrode arrays can also rotate cells. Particles modify the applied fields, allowing electronic analysis of the situation[15]. The small size of dielectrophoresis units is unlikely to allow their use for processing primary suspensions, but the ability to use electronic signal processing and control suggests that they are potentially useful for final stages of microbial separation.

3.6 Immunomagnetic separations

Microbes can be made ferro- or paramagnetic by adsorbing particles of magnetic iron oxides on their surfaces, treating them with Erbium (Er^{3+}), or precipitating ferromagnetic ions on their surfaces. However, more popular methods involve immobilizing them on paramagnetic polystyrene beads (2.8-4.5 μm, Dynabeads, Dynal, UK) or primed, silanized magnetic iron oxide particles (BioMag, Metachem Diagnostics Ltd., UK) by using lectins or antibodies. The attraction of immunomagnetic methods lies in the speed and simplicity with which the target species may be separated by means of a powerful magnet. Techniques can be as simple as collecting a pellet of magnetic cells, washing by resuspending and recollecting, or more sophisticated as in thin-layer magnetophoresis[16, 17]. Separated cells are detected by plating them on normal growth media, electrical impedance methods, or PCR (where immunamagnetic separation can remove inhibitory materials)[18, 19], ELISA[20], or other methods. Immunomagnetic methods can detect dead or severely damaged microbes that are undetectable by standard culture techniques[21].

Separation works best with high concentrations of immunomagnetic particles (10^6 -10^7 particles/ ml for salmonellae and *E. coli* O157[22-24]. Incubations of 10-60 min are required, and attachment increases with time; however, non-specific attachment reduces effectiveness on unduly long incubation. Limitations are the small volume treatable because of the short range of magnetic fields, and a tendency for less-than-quantitative attachment even with great excesses of particle to target cell. A range of techniques and kits based on immuno-magnetic separation (either directly from the initial suspension or after short enrichment) provides earlier detection[25-27]. Commercial magnetic particles primed with lectins or secondary antibodies facilitate the methodology.

3.7 Standing wave ultrasound

Reflection of sound waves in a tube to yield standing waves causes suspended particles to concentrate at nodes, which can then be moved by varying the frequency[28]. It is tempting to think of detaching cells from a test surface, then moving them to a place of collection by the same ultrasonic forces. Particle size, concentration, energy levels, and other factors affect efficiency and it has not yet found practical application in food microbiology.

3.8. Ion-exchange and affinity chromatography

This topic is covered elsewhere and is not discussed here.

4. Preparation of sample suspensions

The mechanism by which the sampling technique detaches microbes from foods is complex and poorly understood. A puzzling feature is that during rinsing, stomaching, blending, etc, the concentration of suspended microbes quickly reaches a plateau; however (where it can be done), repeating the process with fresh diluent liberates more microbes, often at a barely reduced level. If the sampling process is repeated many times the sum of the counts is much higher than a single processing suggests[29-30].

Some authors assume that microbial release follows a second-order reaction, as though microbes in suspension inhibit detachment of others; however, it is difficult to believe that Mass Action has a significant effect. Shaking, swabbing, stomaching or blending all disrupt tissues and perhaps microbes quickly reattach to newly exposed surfaces, or to the sampling device (e.g., swab fibres or stomacher bag). At any rate, one cannot assume that the yield of microbes from a sampling technique (even blending) represents more than a small fraction of their actual level. This consideration is important to rapid pathogen detection methods, since they are compared against traditional pre-enrichment (where single viable cells grow out into a broth and are detected, regardless of their state in the sample). *For samples of low cell concentrations, a sampling technique that does not suspend every pathogen cell will result in poor performance by a rapid detection method.*

Several microbe suspending methods are described below. For "older" methods based on microbial growth it was most important that the technique yielded a maximum level of suspended cells. Faster, more direct detection methods, also demand suspensions that contain a minimum of suspended debris to interfere with the analysis.

4.1. Swabs. Though yielding minimal debris, microbe removal is poorly reproducible and less than needed for quantitative sampling. In the most detailed study, a "wet and dry" swab method gave counts of 1-24% for beef carcass, 27-52% for mutton, 13-67% for pork, and 25-89% for pork belly, compared with counts from excised, blended surfaces[31].

4.2. Improved swabs. The poor performance of swabs may be due to newly exposed surfaces or the swab itself trapping liberated microbes, a result of the high concentration of microbes built up at contact surfaces and an inability to distribute them uniformly through the diluent in the swab. The *Rotorinser* holds a large diluent volume (10 ml for 50 cm^2 area) in a sponge which scrubs the test surface, compressing and rotating to pump liquid around and bring suspended microbes into equilibrium with diluent. The Rotorinser removed more microbes from pork carcass than did excision followed by stomaching (98% at 60 s operation for pork, compared with 86% for a Stomacher)[32].

4.3. Spray methods. Sprays yield low debris levels. A spray method yielded bacteria counts as high as by blender from meats[33]; other workers claimed various success with similar techniques[34]. Difficulty of catching and handling wash waters is a severe limitation.

4.4. Ultrasound and vortex stirring. Insonation gives good microbial removal from hard surfaces. For food samples in glass tubes in an ultrasonic tank, microbe removal compared well with blending and yielded suspensions with very low debris content[35]. Insonation was less effective for prawns which protected surfaces from the energy source, and very poor with comminuted meat. Sonication conditions must be a compromise between effectiveness and lethality from cavitation. Vortex stirring suspended microbes very effectively. Small sample size is a problem for both techniques, and both have lapsed into obscurity.

4.5. Blenders. Although appearing in many standard methods and long assumed to give "quantitative" suspensions of microbes from foods, the need to clean and sterilise after use, overheating, and high levels of suspended debris are severe problems.

4.6. Stomachers. These problems led to development of the Stomacher™ [a] which processes samples in disposable plastic bags. Two paddles reciprocating at 300 rpm crush the sample and drive diluent from side to side in the bag. It causes less tissue disruption than a bladed blender[36]. Early suspicion about this led to many performance evaluations; in

---

[a] STOMACHER™ is a registered trademark of Seward Medical, London, UK

eight of these the Stomacher yielded about the same count as a bladed blender, in six there was no difference, and in about twenty the count was slightly lower. Stomachers replaced bladed blenders in many laboratories, and today at least five "clones" are marketed. They are referred to generically as *paddle-type* processors.

4.7. Pulsifier™. A recent advancement in sample processing, both for traditional microbiological analyses, but particularly for newer detection techniques that are facilitated by "cleaner" starting points, is the Pulsifier[b] (Microgen Bioproducts Ltd., Camberley, Surrey, UK). The Pulsifier also accepts samples in plastic bags. Microbe-suspending energy is applied to the sample bag by a *Beater Ring*, vibrating at 2,900-3,500 rpm. The action is a combination of shock waves and intense agitation. Because it does not crush samples the Pulsifier greatly reduces tissue disruption, even compared to paddle-type processors. An incidental benefit is that hard objects such as rice, bones, and gravel cause less bag-damage. A removable, transparent door allows the action to be viewed and gives excellent accessibility for cleaning.

Since microbes usually exist on surfaces or in easily accessible structures *pulsification* liberates them efficiently. In a comparative trial the average ratio of total aerobic counts *Pulsifier:Stomacher* for 96 samples of representative foods was 1.4[37]. At very low count levels total counts by Pulsifier were approximately twice those obtained by Stomacher, were significantly higher for samples with $<10^5$ CFU/g, and not significantly different for samples with 10^5 - 10^7 CFU/g. As counts at lower levels were obtained on sample suspensions that were not diluted further, the ratios may reflect improved visibility of colonies, or reduced interference by food components, compared with those from the Stomacher. Counts of coliforms, and *E. coli* in *pulsificates* (Fig 1) did not differ significantly from *stomachates* for most food types, were slightly higher for mushrooms, and slightly lower for ground pork[38].

Except with pre-comminuted or powdered foods, pulsified suspensions contained less food debris, although microbial levels were not inferior. For example, for celery and carrot *Pulsifier:Stomacher* total count ratios were 1.3 and 2.5, respectively, but pulsificates were clear while stomachates contained enough debris to interfere with pipetting[37]. Membrane filterability (which deteriorates quickly if there are appreciable levels of suspended particulates) was considerably better for pulsificates (Fig 2). Filtration rate ratios ranged from 1.3x (feta cheese), to 10.7x (broccoli) and 12.2x (beef liver) compared with stomachates. Suspended solids contents ranged from <1% (fresh shrimp) to 102% (oregano and chili powders), and total solids (which include dissolved salts, sugars, etc), ranged from 28% (ocean perch) to 101% (chili powder)[38].

The cleaner suspensions the Pulsifier yields should prove particularly beneficial for techniques in which interference by food components is a problem. For example, improved membrane filterability could improve the limit of detection of DEFT-type tests of pathogens such as the detection of *E. coli* O157:H7 described above[3], to levels close to those of regulatory interest. Reduced debris levels could also be of benefit in polymerase chain reactions, ATP bioluminescence, flow-cytometry and other analyses.

---

[b] PULSIFIER™ is a registered trademark of Filtaflex Ltd., Almonte, Ontario, Canada.

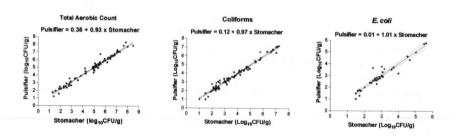

Figure 1.   *Microbial recoveries from foods by Pulsifier and a paddle-type processor.*

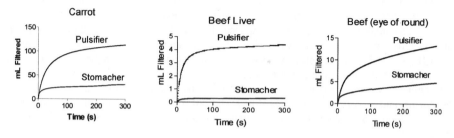

Figure 2.   *Examples of membrane filtration rates of food suspensions prepared by Pulsifier and a paddle-type processor.*

4. REFERENCES

1. A.N. Sharpe, Food Microbiol., 1991, **8**: 167.
2. G.L. Pettipher. *In* Rapid Methods in Food Microbiology, *eds* M.R. Adams and C.F.A. Hope. Progress in Industrial Microbiology, Vol 26, Elsevier, 1989..
3. M.L. Tortorello and D.S. Stewart.. Appl. Environ. Microbiol., 1994, **60,** 3553.
4. P.I., Peterkin, A.N. Sharpe and D.W. Warburton. Appl. Environ. Microbiol., 1982, **43,** 486.
5. A.N. Sharpe, P.I. Peterkin and I. Dudas. Appl. Environ. Microbiol., 1979, 37, 21.
6. A.N. Sharpe and P.I. Peterkin.. Membrane Filter Food Microbiology. Innovation in Microbiology Research Studies Series, Research Studies Press, Letchworth, UK, 1988.
7. U.M. Rodrigues and R.G. Kroll. J. Appl. Bacteriol., 1985, **59,** 493.
8. W. Yamaguchi, J.M. Kopek and A.L. Waldrup. J. Rapid Meth. Autom. Microbiol., 1994, **2,** 287.
9. C.J. Stannard, and J.M. Wood. J. Appl. Bacteriol., 1983, **55,** 429.
10. R.M. Basel, E.R. Richter and G.J. Banwart. Appl. Environ. Microbiol., 1983, **45,** 1156.

11. C.H. Linhardt, 1987. J. Appl. Bacteriol., 63(6): XXIII. (Conference paper).
12. R.V. Sharma, R.T. Edwards and R. Beckett. Appl. Environ. Microbiol., 1993, **59**, 1864.
13. R.P. Betts, *in* Rapid Methods and Automation in Microbiology and Automation, *eds* R.C. Spencer, E.P. Wright and S.W.B. Newsom, Athenaeum Press, UK, 1993.
14. O. Stendahl, K-E. Magnusson, C. Cunningham and R. Edebo. Infect. Immun. 1973, **7**, 573.
15. Y. Huang, Holzel, R., Pethig, R., and Wang, X.B. 1992. Phys. Med. Biol., 1992, **37** 1499.
16. M.J. Payne, S. Campbell, R.A. Patchett and R.G. Kroll. J. Appl. Bacteriol., 1992, **73**, 41.
17. I. Safarik, M. Safariková and S.J. Forsythe. J. Appl. Bacteriol., 1995, **78**, 575.
18. A.C.Fluit, R. Torensma, M.J. Visser, et. al... Appl. Environ. Microbiol., 1993, **59**, 1289.
19. M.N.Widjojoatmodjo, A.C. Fluit, R. Torensma, B.H. Keller and J. Verhoef. Eur. J. Clin. Microbiol. Infect. Dis., 1991, **10**, 935.
20. L. Krusell and N. Skovgaard. Int. J. Food Microbiol., 1993, **20**, 123.
21. L.P. Mansfield and S.J. Forsythe. Lett. Appl. Microbiol., 1993, **16** 122.
22. E. Skjerve, L.M. Rorvik and O. Olsvik. Appl. Environ. Microbiol., 1990, **56**, 3478.
23. A.E.M. Vermunt, A.A.J.M. Franken and R.R. Beumer. J. Appl. Bacteriol., 1992, **72**, 112.
24. P.M. Framatico, F.J. Schultz and R.L. Buchanan. Food Microbiol., 1992, **9**, 105.
25. P.A. Chapman and C.A. Siddons. Food Microbiol., 1996, **13**, 175.
26. J.M.C. Luk and A.A. Lindberg. J. Immunol. Methods, 1991, **137**, 1.
27. T. Tomoyasu. Appl. Environ. Microbiol., 1992, **58**, 2679.
28. C.A. Miles, M.J. Morley, W.R. Hudson and B.M. Mackey. J. Appl. Bacteriol., 1995, **78**, 47.
29. P.B. Price. J. Inf. Dis., 1938, **63**, 301.
30. H.S. Lillard. J. Food Prot., 1988, **51**, 405.
31. M. Ingram and T.A. Roberts. Roy. Soc. Health J., 1976, **96**, 270.
32. A.N. Sharpe, C.I. Bin Kingombe, P. Watney, L.J. Parrington and I. Dudas. J. Food Prot., 1996, **59**, 757.
33. D.S. Clark. Can. J. Microbiol., 1963, **11**, 407.
34. G. Reuter, D. Sasse and G. Sibomana. Arch. Lebensmitt., 1979, **30**: 126.
35. A.N. Sharpe and D.C. Kilsby. J. Appl. Bacteriol., 1970, **33**, 351.
36. A.N. Sharpe and A.K. Jackson. Appl. Microbiol., 1972, **24**, 175.
37. Fung, D.Y.C., A.N. Sharpe, B.C. Hart and Y. Liu. 1998. J. Rapid Meth. Autom. Microbiol., 6: 43-49.
38. A.N. Sharpe, E.M. Hearn and J. Kovacs-Nolan. Membrane Filtration Rates and HGMF Coliform and *Escherichia coli* Counts in Food Suspensions by Pulsifier™. (*submitted for publication*).

THE SEPARATION OF FOOD PATHOGENS USING CHROMATOGRAPHIC TECHNIQUES

T. N. Whitmore and P.D. Gray

Microbiology Group
WRc plc
Marlow
Buckinghamshire SL7 2HD

1 INTRODUCTION

The selective separation of the target bacterial cell from food enrichment cultures should improve the reliability of detection. The separation of *Escherichia coli* from mixed cell suspensions has been demonstrated using ion exchange chromatography (IEC)[1] and hydrophobic interaction chromatography (HIC)[2]. This paper describes the application of liquid chromatographic techniques to separate binary combinations of bacteria and foodborne bacterial pathogens from inoculated food samples.

2 MATERIALS AND METHODS

2.1 Test organisms

The *E. coli* O157:H7 strains (NCTC 12900 and 12079) and *E. coli* K88a (NCTC 10650) were obtained from the NCTC, Colindale, London; and a clinical *E. coli* O157:H7 strain (30-2C4) from Dr. R.W.A. Park (University of Reading).

2.2 Chromatographic media

The hydrophobic interaction and mimetic ligand (ML) chromatography media were obtained from Affinity Chromatography Ltd, Girton, Cambridge, U.K. and the Toyopearl™ ion exchange media from TosoHaas, Linton, Cambridge, U.K.

2.3 Chromatographic conditions

The HIC and ML columns containing 1 ml bed volume of media were of 0.8 cm internal diameter and were operated under gravity flow at a rate of 0.2 ml min[-1] equivalent to a linear flow rate of 0.4 cm min[-1]. The support matrix for the HIC and ML media was 6% cross-linked agarose. The IEC columns of internal diameter 1.5 cm contained 5 ml bed volume of coarse grade (100 μm) Toyopearl™ resin were operated under gravity flow at a flow rate of 0.9 ml min[-1] equivalent to a linear flow of 0.5 ml min[-1].

2.4 Growth media

The *E. coli* serotypes were cultivated overnight at 37 °C in nutrient broth (Oxoid CM1) and enumerated on MacConkey Agar (Oxoid CM7) at 37 °C using the spread plate technique or, when discrimination between the two serotypes was required, sorbitol MacConkey agar (Oxoid CM 813).

3 CHROMATOGRAPHIC SEPARATION

3.1 Hydrophobic Interaction Chromatography

3.1.1 Hydrophobic properties of test bacteria. The retention of the test strains (1×10^6 - 1×10^7 colony forming units) suspended in 50 mM phosphate buffer containing 1 M ammonium sulphate to phenyl-agarose demonstrated the low affinity of each of the *E. coli* O157:H7 test strains compared with the K88a serotype.

3.2.2 Separation of bacteria using HIC. The partial separation of an *E. coli* O157:H7 serotype (NCTC 12900) from the K88a serotype by exploiting the difference in affinity to phenyl agarose under high salt conditions is shown in Figure 1. The proportion of the O157:H7 serotype increased from 35% in the column influent to 99.5% in the high salt (1 M ammonium sulphate) eluate, equivalent to an enrichment ratio of 335:1. The second low salt (50 mM phosphate) eluate fraction was slightly enriched in the K88a serotype.

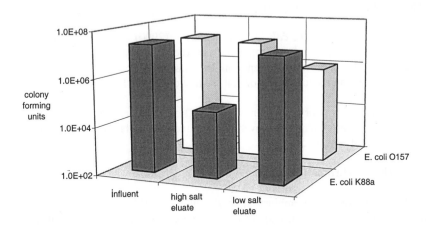

Figure 1. *Separation of E. coli serotypes using hydrophobic interaction chromatography*

Further experiments indicated that the ammonium sulphate concentration could be reduced to 0.5 M with no loss of separation efficiency.

The selective recovery of *E. coli* O157:H7 from spiked minced beef enrichment cultures using modified EC³ (mEC) and nutrient broth (NB) media was demonstrated using HIC (Figure 2). The proportion of *E. coli* O157:H7 compared with non-sorbitol fermenting organisms, enumerated using sorbitol MacConkey agar, increased by a ratio of 650:1 and at least 350:1 for the nutrient broth and mEC enrichment media respectively.

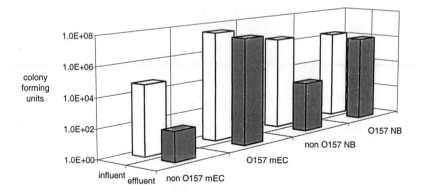

Figure 2. *Separation of E.coli O157:H7 (NCTC 12900) from food enrichment cultures using hydrophobic interaction chromatography*

Non-*E. coli* O157:H7 cells were not detected in the column effluent from the mEC enrichment. The detection limit of 100 colony forming units is shown (Figure 2).

3.2 Mimetic ligand chromatography

The low affinity of *E. coli* O157:H7 (NCTC 12900) compared with *E. coli* K88a to mimetic ligand chromatographic adsorbents was exploited to separate a combination of the organisms. An enrichment ratio of $7.5 \times 10^4 : 1$ was obtained following chromatographic separation on Mimetic Green 1 agarose in 25 mM phosphate buffer.

A comparatively poor separation of *E. coli* O157:H7 from a spiked minced beef enrichment broth was obtained, however, with an increase in the ratio of this serotype to non-O157:H7 strains of 5:1.

3.3 Ion exchange chromatography

The separation of a combination of *E. coli* O157:H7 and *E. coli* K88a was unsuccessful using both weak and strong cation exchange resins. The weak anion exchanger (DEAE-650C) gave a good resolution which was superior to the strong anion exchangers QAE-550C and Super-Q 650C. The low affinity of the O157:H7 serotype to the column matrix resulted in a significant enrichment (at least 8000 :1) of this serotype in 10 mM, pH 6 phosphate buffer (Figure 3.). Increasing the salt (sodium chloride) concentration in the elution buffer to 1 M caused some elution of the K88a serotype but approximately 98% of the applied cells remained within the column matrix.

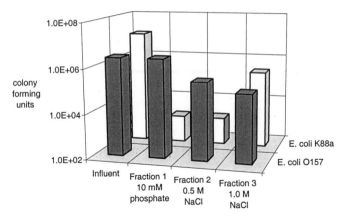

Figure 3. *Separation of E. coli serotypes using anion exchange chromatography*

E. coli K88a was not detected in fractions 1 and 2. The threshold of detection of 1.6×10^3 colony forming units is shown (Figure 3).

4 CONCLUSIONS

The potential of three chromatographic techniques to separate binary combinations of bacterial populations and *E. coli* O157:H7 from inoculated food enrichments was demonstrated. Each of the techniques gave an efficient separation of two serotypes of *E. coli* using isocratic elution, in which the non-pathogenic K88a serotype was preferentially retarded. The low affinity of *E. coli* O157:H7 for hydrophobic and mimetic ligand media may be useful properties of this pathogen to exploit for separation from a variety of matrices. However, ion exchange chromatography may be a more versatile technique to separate a wider range of target micro-organisms.

Acknowledgement

The study was supported by the U.K. Ministry of Agriculture Fisheries and Food.

References

1.   J.M. Wood and P.A. Gibbs, 'Developments in Food Microbiology', Daniels, R. Ed. Applied Science Publishers, London, 1982. p.183.

2.   P.D. Patel and J.M.Wood, 'Rapid Methods and Automation in Microbiology and Immunology', Habermehl, K.-O. Ed. Springer Verlag. New York, 1984. p. 665.

3.   A.R. Bennett, S. MacPhee and, R.P. Betts. *Lett. Appl. Micro.* 1996, **22,** 237.

THE ACCREDITATION OF MICROBIOLOGICAL TEST METHODS

Roy Betts

Campden & Chorleywood Food Research Association, Chipping Campden, Gloucestershire, GL55 6LD, UK

1 INTRODUCTION

Historically, microbiological test methods have been based on so called conventional techniques in which microorganisms are cultured in broths and on agars, and are generally identified by their morphology (either colonial or cell) and/or biochemistry. Looking back at this historical perspective, methods tended to be developed through research done by recognised independent experts, followed by suitable peer review. Following this, such new methods could become very widely accepted reference methods.

More recently, there has been an accelerated search for new test methods driven by the needs of testing laboratories who want faster, more automated, user friendly methods with more objective end points. This challenge to devise new methods has been taken up by commercial method manufacturers, who have developed a wide range of rapid and automated microbiological test methods.

To be of value to a laboratory, a rapid microbiological method must be proven to have defined parameters of operation (sensitivity, specificity, selectivity) and to be applicable to a variety of sample types and laboratory settings. The task of validation of new test methods is, therefore, of great importance, as it provides evidence that a new method does work when tested under defined conditions and gives end users confidence in results obtained.

A further driver to the use of method validation procedures is the increasing requirement for laboratory accreditation. Accredited laboratories must use methods that have been subjected to a validation, thus the progression towards the use of formal method validation schemes is enhanced.

The terminology used by various people for the testing of an analytical method can be confusing. The terms validation, evaluation, assessment and certification are often used interchangeably; however, they do have subtly different meanings (Chambers English Dictionary):

Validate: to be founded in truth, fulfilling all necessary conditions
Evaluate: to determine the value of
Assess: to value, to estimate, a valuation
Certify: to declare formally or in writing, with authority

It is important to try to use the correct term when referring to a method test procedure. It is also important to distinguish method testing depending on who organises the tests or reports on the work. If testing is done by the manufacturer, it could be considered to be first party testing. If it is done by a user laboratory, it would be second party testing, and if it were done by an independent organisation then this is third party testing.

Second party testing is a critical part of assessing the suitability of a method in a user laboratory; it tests the individual laboratory's capacity to operate the method on its typical sample types. This type of testing is, however, only relevant to the individual laboratory undertaking the work. Independent third party testing will satisfy a wide range of potential users that a method is fit for purpose. It is not laboratory specific and will satisfy accreditation bodies that a new method will give equivalent results to a recognised reference method.

In an ideal world, there would be a single method validation scheme that was acceptable throughout the world. This would enable method manufacturers to put their time and money into one scheme and, once validation was achieved, to use this to sell their products world-wide. Unfortunately, this has not occurred and a variety of method validation schemes have been developed. This has given method manufacturers the difficulty of not knowing if they need their kit tested by one or all of the systems and kit users the problem of not knowing what all the validation systems mean.

This paper will briefly review the major method validation schemes in order to give the reader an idea of how each operates, what technical data is needed, or generated, and where the results of the schemes are accepted.

The validation schemes described are:

- AOAC International: Official Methods Programme
- AOAC International Peer Verified Methods
- AOAC Research Institute Performance Tested System
- Association Français de Normalisation (AFNOR)
- MICROVAL
- DANVAL (NORVAL)
- European Microbiological Method Assessment Scheme (EMMAS)

1.1 AOAC International Official Methods Programme

The Association of Official Analytical Chemists (AOAC) was founded in the late 1800's, essentially to ensure that chemical test methods were standardised. The organisation now considers all types of chemical and microbiological test methods, with a primary objective of ensuring high standards of analysis in testing laboratories.

The AOAC International Official Methods Programme is used to validate all non-proprietary (i.e. not kit-based) methods. It is based on the use of a pre-collaborative study, followed by a collaborative study, then a final report of the studies with a recommendation that the method should (or should not) be adopted. Each study is overseen by an AOAC Associate Referee.

The pre-collaborative study is conducted in a single laboratory, and is used to demonstrate that the method works with a variety of sample types; it also helps resolve any problems with the method before the collaborative study. Once the pre-collaborative study has been done and approved by the methods committee, then the collaborative study is started. The collaborative study is a large multi-sample, multi-laboratory test of the method. It involves between eight and fifteen laboratories testing a number of samples of various foods inoculated with test microorganisms. In the case of quantitative methods, a minimum of eight laboratories must each test five samples in duplicate. For qualitative tests, a minimum of 15 laboratories must test five samples at two levels per matrix type, as well as five negative controls.

When the collaborative study is complete, the Associate Referee in charge of the study prepares a report for publication. If the collaborative study was considered a success, then the method is adopted by the AOAC as a First Action Method and is incorporated into the official methods of analysis. At the third annual AOAC meeting after the method has been adopted as First Action, it may be considered for adoption as a Final Action method, subject to any negative feedback that may have been received.

1.2 AOAC Peer Verified Methods

The AOAC Peer Verified Methods programme is intended to provide a class of tested non-proprietary methods which have not been subjected to a full collaborative study. The stages involved in a peer verified method test are:

(1)   establishment of acceptable performance parameters within a laboratory; and

(2)   demonstration of acceptable performance in a second or third laboratory.

The procedure is managed by a method author who is responsible for developing the testing criteria and protocol and for recruiting the test laboratories. After testing, the results are passed to an AOAC Technical Referee who seeks a review by qualified experts. The referee, on the basis of a minimum of two positive reviews, grants AOAC Peer Verified status to the method.

1.3 AOAC Research Institute Performance Tested Programme

The Performance Tested Programme is intended for proprietary (i.e. kit-based) methods. It was designed to be a fairly rapid approach to obtaining validation, and is based on tests done in a single laboratory to assess if a method performs according to the manufacturer's claims.

The kit manufacturer supplies the AOAC RI with information on the test method. This is assessed by two expert reviewers who devise a test protocol to test the manufacturer's claims. A single expert laboratory undertakes the practical work, and provides results to the reviewers who decide if the method operates according to the manufacturer's instructions. If the latter is the case, then the method is granted AOAC RI Performance Tested Status. The method is reviewed on a yearly basis to assess if it has changed in any way and requires retesting.

1.3.1 Integration of the AOAC Programmes. Recently, there have been moves within the AOAC to integrate the three programmes as it was found that users were confused as to the meaning and requirements of the three. A diagram of the operation of the integrated system is shown opposite.

AOAC/AOAC RI

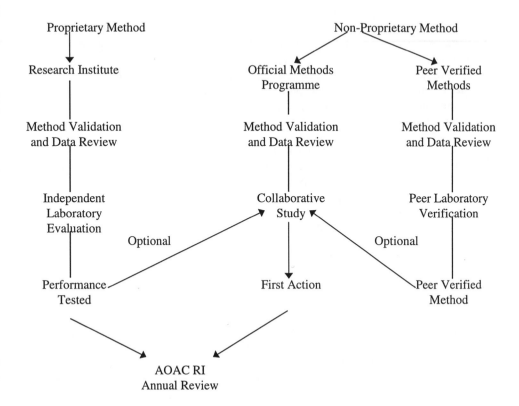

1.4 AFNOR

The AFNOR system for the validation of rapid test kits was set up in 1989, in order to test that new methods gave results that were equivalent to reference methods.

Rapid methods are validated on the basis of a preliminary study, a collaborative study and an audit. The aim of the preliminary study is to compare the performance of a rapid method with the reference method, and is done by an expert laboratory. A report of this work is prepared and presented to the AFNOR Technical Board who advise if the method is equivalent to the reference method. Once this occurs, the collaborative study can begin. This is organised by the expert laboratory, and involves a minimum of six laboratories who receive samples provided by the expert laboratory. Results are provided to the AFNOR Technical Board who advise whether or not the method should be validated.

Parallel to the collaborative studies, AFNOR conducts an audit of the manufacturer to ensure that they have a quality system in place which is equivalent to ISO 9002, as well as requirements concerning in-house control of products.

AFNOR additionally organises surveillance of the method and the manufacturer at two yearly intervals. After four years, if AFNOR validation is to be renewed, then a further collaborative study and quality audit is required.

1.5 MICROVAL

MICROVAL was developed because of a lack of an acceptable microbiological method validation scheme recognised throughout Europe. MICROVAL is a project designed to devise a European method validation/certification scheme and has involved 21 partners from seven different EU member states. The partners have been meeting over the last four years to devise a set of operating rules for a certification scheme, and a set of technical rules for method validation. The procedure which has been developed involves two parts: a comparative study done in a single expert laboratory; and a collaborative study done in a number of laboratories using samples supplied by the expert laboratory.

The comparative study is done to enable the test's specificity, limit of detection, relative accuracy, sensitivity and linearity to be assessed. The collaborative study, done in a minimum of eight laboratories, allows determination of precision, repeatability and reproducibility. All results will be assessed by two expert reviewers who recommend or reject the test. Final acceptance is from a MICROVAL General Committee who arrange for method certification through a suitable certification body.

At present, the MICROVAL scheme is not operational. The technical rules are being considered by a CEN committee for adoption as a CEN standard for the validation of microbiological test methods.

1.6 DANVAL

DANVAL is a system organised by the Danish Government (Danish Veterinary Services, National Food Agency and Fisheries Department) for the validation of microbiological test methods. It is based on the testing of a number of categories of contaminated food product using a new test kit and a reference method. The procedure was originally defined for *Salmonella* tests and thus the categories of food were chosen to reflect those likely to be tested for *Salmonella*. *Listeria* tests have now also been covered.

In the future it is possible that a DANVAL type scheme could be adopted more widely by Nordic countries in a scheme known as NORVAL.

1.7 EMMAS

The European Microbiological Method Assessment Scheme was developed by the two UK based Food Research Associations (Campden & Chorleywood Food Research Association and Leatherhead Food Research Association) at the request of their industrially based membership. It was designed to be an assessment scheme giving the end user all of the information they require to make an effective choice of method. Therefore, as well as defining the technical parameters of the method, such as sensitivity, specificity, etc., the scheme would also give the end user details of test rapidity, equipment required, costs and sample throughput per analyst per day. The user could therefore make an effective decision as to whether the method could fit into their laboratory's working environment.

The EMMAS procedure requires initial testing of the method to establish sensitivity, specificity and ability of the test to recover the target analyte from a range of food types. This allows the basic parameters of the test to be established. After this, the test undergoes a ring trial in a minimum of six end-user laboratories. There it is tested under 'real' conditions with real food samples; the results are returned to one of the Research Association for analysis. A test report can then be written which includes all details of the kit useful to a potential user. Within EMMAS, if previously generated results for a test method are available, then these can be used within the evaluation, reducing the requirement to generate 'new' data.

2 CONCLUSIONS

This paper gives a short report on microbial method validation and the variety of method validation schemes currently available and operating throughout the world. There are no recognised schemes operating throughout the Pacific Rim countries; here, AOAC systems tend to be recognised. It is, however, apparent that Australia is carefully considering the introduction of a validation scheme.

New test methods can offer the food industry advantages over older methodologies. Examples include greater rapidity, higher sample throughput, simpler operating procedures and greater automation. Validation will allow laboratories to use such methods with confidence and this will add to safety and quality attributes of manufactured foods. It is, however, important that the different validation schemes begin to move together, and perhaps have a degree of mutual acceptability. This will make them easier to understand and more cost effective for method manufacturers to use.

THE USE OF A MINIMAL NUMBER OF VAPOUR SENSORS FOR THE ASSESSMENT OF FOOD QUALITY

B. P. J. de Lacy Costello, R. J. Ewen, H. E. Gunson, N. M. Ratcliffe* and
P. T. N. Spencer-Phillips

Faculty of Applied Sciences,
University of the West of England,
Coldharbour Lane, Bristol, BS16 1QY.

1 INTRODUCTION

The action of micro-organisms on a variety of foodstuffs is known to release a range of volatile organic compounds (VOCs) which constitute the characteristic malodours associated with the spoilage process. In a recent study[1] we found that a general increase in the concentration of VOCs served as the best marker for *Erwinia carotovora* infections of potato tubers. *E. carotovora* is the primary cause of soft rot and is a major problem in the bulk storage of many vegetable crops, especially potato tubers. A sensor system to monitor the increase in VOCs could be used for the early detection of *Erwinia* infections in potato stores.

The use of chemical sensors has been widely reported[2] for the assessment of food quality. Often the systems used in these applications combine large numbers of sensors in an array.

In previous work[3] the use of ceramic sensors was investigated for the detection of *Penicillium* infections in stored oranges. It was also important to detect damage to the oranges as this increased their susceptibility to fungal infection. It was found that to detect damage to the outer skin of an orange only one sensor which was sensitive to limonene vapour was required.

Work was also undertaken[4] to produce a system capable of detecting off-flavours in air-cured Iberian hams. Many sensor types were utilised in the developmental stages but it was found that the majority did not exhibit the required sensitivity or high stability characteristics necessary for incorporation into a working device. The final device incorporated two ceramic sensors and one electrochemical sensor and was capable of distinguishing between hams with and without off flavours. In trials in a factory environment the sensing head was attached to a mechatronics system and was capable of classifying a group of 400 hams (360 good and 40 bad).

In this work we report the fabrication and testing of a system incorporating three ceramic type sensors which is capable of detecting one *E. carotovora* infected potato tuber amongst 100 kg of sound tubers. The device we report is portable and could be of use in the monitoring of potato stores for the onset of disease.

2 EXPERIMENTAL

2.1 Fabrication of Sensors

A more detailed description of the production of these sensor types is reported elsewhere[4].

2.1.1 Evaporated film sensors. Tin metal was evaporated through an oxygen plasma giving films of tin dioxide at the electrode surface. Films of 150 nm and 940 nm were fabricated and used in the device.

2.1.2 Thick film sensors. Tin oxide and zinc oxide composite sensors (50/50 m/m) were fabricated by grinding the powders in water and depositing the resulting paste across interdigitated electrodes.

2.2 Fabrication of Sensor-array System

The three sensors were mounted in a chamber of 0.1 l and a constant flow (0.3 l/min) of air was maintained across the sensors using a low voltage diaphragm pump. The inlet was connected to a 1 m length of plastic tubing (4 mm internal diameter) which allowed sampling at remote distances from the system. The system was powered by a sealed lead acid battery. The resistance and operating temperature (sensors were maintained at 350°C) of each sensor were displayed on an LCD mounted on the outer casing.

The sensor system can be interfaced to a personal computer allowing the pass/fail and system settling criteria to be set and downloaded to the system. Once the values have been downloaded it is not necessary to maintain the computer link to operate the system. In the case of settling criteria the drift of the sensors with time is monitored and tested against the set values for each sensor.

2.3 Method Used for the Inoculation of Potato Tubers with *Erwinia carotovora*

Potato tubers were surface sterilised and inoculated with a bacterial cell suspension of *E. carotovora subsp. carotovora.* The inoculated tubers were then incubated at 20°C for 7 days before being used in the sensing trials. A more detailed account of the methodology used for the inoculation of tubers is reported elsewhere[1].

2.4 Testing of Sensor Systems to *Erwinia* Infected Potatoes Amongst Fresh Tubers

Testing of sensors and prototype sensor systems to pure vapours and to *Erwinia* infected and sound potato tubers has been undertaken.

The prototype device was tested against one *Erwinia* infected tuber (0.11 kg) in a quickfit jar (2 l) containing 9 uninfected tubers (1 kg). The air was sampled for 60 seconds and after this time the change in each sensor was recorded and the responses of the three sensors were summed to give a system output. For comparison an identical jar containing sound tubers was tested alternately. A sound/infected threshold based on the results of preliminary tests was set. The results of this analysis are displayed in Figure 1.

A simulated storage crate containing 100 kg of tubers was used in the subsequent tests. The headspace at the top surface of the potato tubers was sampled (60 s) at 5 different points. The first sampling point was at the centre and the other four points were at the half way points between the centre and the outer corner of the crate. An *Erwinia* infected tuber was buried at the bottom of the crate as close to the centre as possible and the tests were repeated. The results are displayed in Figure 2.

3 RESULTS AND DISCUSSION

Figure 1 shows that the system was able to detect the presence of infected tubers amongst sound tubers with only one misclassification in 128 tests. The proportion of infected material is relatively high and good separation of the data was observed. In a useful system a few infected tubers would have to be detected in a crate containing 1000 kg of sound tubers.

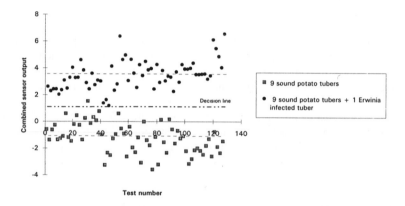

Figure 1 *The output from the sensor system when exposed to sound tubers and subsequently with 1 Erwinia infected tuber added*

The results displayed in Figure 2 show that the system was åble to detect the presence of one infected tuber amongst 100 kg of sound tubers. The sampling point clearly had an effect on the discriminating ability of the system. Sampling directly above the source of infection (point 1) yielded the greatest discrimination but the system was still able to detect the presence of the infected tuber even when sampling at points 2, 3, 4 and 5 an appreciable distance from the source. It is apparent from the results that the thick film sensor may give sufficient discrimination to allow a system based on one sensor to be considered.

The sensors were affected by changes in humidity and by interfering volatiles in the atmosphere causing relatively large fluctuations in resistance. By controlling the humidity the stability of the system was improved whilst the sensitivity remained relatively unchanged. The stability is important as it limits the discriminating ability of the system. Thus, the use of this method enables a better separation of the infected and non-infected data allowing a reliable pass/fail criterion to be set.

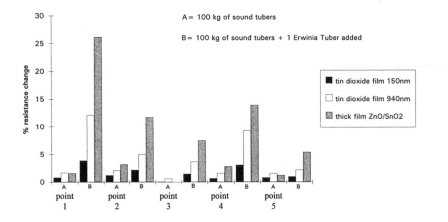

Figure 2 *Response of individual sensors within the system to 100 kg of sound tubers and subsequently with 1 Erwinia infected tuber added.*

4 CONCLUSIONS

Sensors that are highly sensitive to the vapour emitted by *Erwinia* infected potato tubers have been produced. A prototype system incorporating three ceramic sensors has been fabricated which is capable of detecting one *Erwinia* infected tuber amongst 100 kg of sound tubers under simulated storage conditions, equivalent to 10 infected tubers per standard 1000 kg crate. The next stage is to test the system at a potato storage site.

References

1. B. P. J. de Lacy Costello, P. Evans, R. J. Ewen, H. E. Gunson, N. M. Ratcliffe and P. T. N. Spencer-Phillips, *Plant Pathol.*, 1999, in press.
2. J. Brezmes, B. Ferreras, E. Llobet, X. Vilanova and X. Correig, *Anal. Chim. Acta*, 1997, **348**, 503.
3. D. C. Cowell, R. J. Ewen, C. E. Flynn, B. Goldie, J. P. Hart, S. J. Hawkins, T. J. A. R. Hitch, C. L. Honeybourne, D. V. McCalley and N. M. Ratcliffe, *Life Chem. Rep.* 1994, 11, 333.
4. A. K. Abass, B. P. J. de Lacy Costello, P. Evans, R. J. Ewen, J. P. Hart, N. M. Ratcliffe and R. K. M. Wat. 1998 Proceedings of the 5th International Symposium on Olfaction and Electronic Noses.

Acknowledgements

The authors would like to thank the British Potato Council for funding this project.

RAPID DETECTION OF ENTEROBACTERIACEAE IN DEHYDRATED FOODS USING IMPEDANCE MICROBIOLOGY: THE CRITICAL ROLE OF THE PRE-ENRICHMENT PHASE

A Pridmore and P Silley

Don Whitley Scientific Limited
14 Otley Road
Shipley BD17 7SE
UK

1 INTRODUCTION

The bacteriological examination of food in order to monitor sanitary practices, processing efficacy and post processing contamination encompasses screening for specific groups of bacteria or individual species. These organisms are commonly referred to as indicator organisms. Particular groups or genera may also be used to assess the potential risk of closely related pathogens being present in foods, a function for which the term index organism has been proposed[1,2].

The most popular group of indicator organisms has been the coliforms, although this group does not have great specificity. The genera detected by standard coliform tests are those members of the Enterobacteriaceae that are able to utilise lactose with acid and gas production. Thus, slow lactose-fermenting or lactose negative organisms would be overlooked. To improve this situation with respect to assessment of food safety and quality, it is now common practice to use a total Enterobacteriaceae count instead[3,4].

The present study examined the detection of Enterobacteriaceae in spray-dried baby foods produced from rice flour. This raw ingredient provides a natural source of Enterobacteriaceae, the levels of which are reduced by consecutive heating and drying stages in the manufacturing process. Detection of the surviving sub-lethally damaged organisms had conventionally been achieved using a two-stage enrichment/confirmation procedure. This provided adequate sensitivity for the target organisms but was labour-intensive and had a 48 hour test duration. Positive release of product on the basis of Enterobacteriaceae levels could potentially be accelerated by the use of a rapid detection assay. Work therefore commenced to develop an enhanced detection protocol for this group of organisms using the Rapid Automated Bacterial Impedance Technique (RABIT).

2   MATERIALS AND METHODS

2.1   Test Samples

Both savoury and dessert varieties of rice flour-based baby food were used to evaluate the detection methods used in this study.   Each was naturally contaminated with Enterobacteriaceae at a level of approximately 0.5 cfu g^{-1}.   These organisms had originated from untreated rice flour and had survived the normal manufacturing process.

2.2   Conventional Methodology

An initial tenfold dilution of each sample was prepared by rehydrating the required mass with Maximum Recovery Diluent (MRD) and processing in a Stomacher for 30 seconds.   Rehydrated samples were allowed to stand at room temperature for 30 minutes and 100 x 10 ml aliquots were transferred to equal volumes of double-strength Enterobacteriaceae Enrichment Broth (EE Broth - Oxoid Limited; CM317).   These were incubated at 30°C for 24 h to promote selective recovery of the target organisms.   Each broth was then used to stab inoculate a 10 ml tube of Violet Red Bile Glucose Agar (VRBGA - Oxoid Limited; CM485) which was incubated at 30°C for a further 24 h. Tubes showing a colour change from maroon to bright purple in the lower parts of the agar were counted and were regarded as presumptive positive Enterobacteriaceae tests.

2.3   RABIT Methodology

2.3.1 Principle.   Detection of microbial growth using the RABIT system is achieved by adding the prepared test sample to a suitable culture medium in reusable impedance cells (working volume 2 ml to 10 ml).   Impedance cells are inserted in the RABIT incubator modules which maintain the selected incubation temperature while recording conductance measurements across the cell contents.   Microbial utilisation of the culture medium during growth produces conductance changes which are presented graphically by the accompanying software.   A positive result is recorded automatically when the rate of conductance change exceeds a threshold selected by the operator.   In the present study, all RABIT protocols were tested in parallel with the conventional methodology using an equal number of replicates.

2.3.2 Single Stage RABIT Procedure.   An initial tenfold dilution of each test sample was prepared in MRD as described above.   After standing for 30 minutes, 100 x 5 ml aliquots were transferred to equal volumes of impedance medium 1 in sterile RABIT cells.   The medium consisted of Whitley Impedance Broth (WIB - Don Whitley Scientific Limited; G50001) supplemented with 0.5 g sodium lauryl sulphate per litre and was prepared at double strength.   RABIT cells were connected to the system and incubated at 30°C for 24 h, during which a positive test result was recorded if the rate of conductance change exceeded 10 μS in three successive 6 minute intervals.   To provide confirmation of RABIT test results, the contents of each incubated impedance cell were used to stab inoculate a 10 ml tube of VRBGA, which was incubated and interpreted as described previously.

2.3.3 RABIT Procedures with Pre-enrichment. Three pre-enrichment protocols were evaluated in conjunction with impedance detection. Initial tenfold dilutions of each test sample were prepared in *(a)* Buffered Peptone Water (BPW) *(b)* BPW + 30 mg Brilliant Green per litre and *(c)* BPW + 20 mg novobiocin per litre. Each preparation was incubated at 30°C for 2 h and 30 x 5 ml aliquots were transferred to equal volumes of impedance medium 2 in sterile RABIT cells. The medium consisted of WIB + 0.5 g sodium deoxycholate per litre and was prepared at double strength. RABIT cells were incubated as described above. All BPW preparations were re-incubated at 30°C for a further 2 h and the RABIT tests were repeated. The results of all RABIT tests were again subjected to confirmation by subculture into VRBGA.

3 RESULTS

Using the single stage RABIT method a high rate of detection was reported by the system for both product types (Table 1a). However, the frequency of "false positive" detection (positive RABIT tests not confirmed by a VRBGA colour change) was up to 20% while the number of confirmed positive tests represented a lower rate of Enterobacteriaceae detection than that achieved using the conventional methodology.

A 4 h pre-enrichment phase in unmodified BPW increased the number of confirmed positive results in the subsequent RABIT tests (Table 1b). In this case the rate of Enterobacteriaceae detection was almost double that achieved conventionally, but the rate of "false positive" detection increased to 50% of replicate tests. A 4 h pre-enrichment phase in BPW + Brilliant Green produced the best correlation between conventional and RABIT methodologies (Table 1c). The impedance method produced a higher rate of confirmed positive results while "false positive" detection was observed with a mean frequency of only 1%. Pre-enrichment in BPW + novobiocin prior to RABIT testing produced distinct results for the two product types (Table 1d): in savoury product the detection rates of the two methodologies were very similar, while in dessert product the RABIT method exhibited a detection rate for the target organisms more than three times that of the conventional method.

The reduced pre-enrichment of 2 h duration prior to RABIT testing produced unsatisfactory results (data not shown). Using the optimal BPW + Brilliant Green protocol the confirmed positive detection rate was below that of the conventional methodology while the frequency of "false positive" detection was approximately 5%.

4 DISCUSSION

The efficacy of impedance instrumentation for detecting sub-lethally damaged bacteria has previously been demonstrated for a variety of sample types[5,6,7]. It should be appreciated, however, that this technology is dependent on active bacterial metabolism, for which adequate resuscitation of the target organism is a prerequisite. In the initial

Table 1   *Comparison of conventional and RABIT methodologies for detection of Enterobacteriaceae in spray-dried baby food*

(a)   Single stage RABIT test

Product type	Savoury	Dessert
Number of replicates	100	100
Total positive RABIT tests	54	94
Confirmed positive RABIT tests	41	74
Confirmed positive - conventional method	73	68

(b)   RABIT tests after BPW pre-enrichment

Product type	Dessert
Number of replicates	30
Total positive RABIT tests	28
Confirmed positive RABIT tests	13
Confirmed positive - conventional method	7

(c)   RABIT tests after BPW + Brilliant Green pre-enrichment

Product type	Savoury	Dessert
Number of replicates	30	30
Total positive RABIT tests	9	9
Confirmed positive RABIT tests	7	9
Confirmed positive - conventional method	3	3

(d)   RABIT tests after BPW + Novobiocin pre-enrichment

Product type	Savoury	Dessert
Number of replicates	30	30
Total positive RABIT tests	6	25
Confirmed positive RABIT tests	4	25
Confirmed positive - conventional method	3	

phase of the present study the Enterobacteriaceae within the test products did not regain sufficient metabolic vigour to allow detection within a 24 h RABIT test alone: competing organisms were thus detected at the expense of the target species. In products which have undergone heating, dehydration or a combination of these processes the predominant surviving bacteria are commonly gram positive sporeformers, the growth rate of which is likely to exceed that of sublethally damaged gram negative species upon entry into a high-moisture nutrient matrix.

In this scenario the pre-enrichment phase has a dual function: resuscitation of the target species must be accompanied by selective inhibition of the competing microflora. Prudent selection of the inhibitory agent ensures that the latter process occurs relatively quickly following rehydration of the test product, promoting multiplication of the target species in the absence of competitive inhibition. In the present study, brilliant green dye and novobiocin were selected as specific inhibitors of gram positive species. The design of the test protocol caused the concentration of each agent to be halved when the pre-enrichment broth was added to the impedance medium. This permitted the use of a relatively high inhibitor concentration in conjunction with a short pre-enrichment; when combined with RABIT detection the total test duration was 28 h. In practice, the majority of sample replicates were detected within 20 h of RABIT test initiation, thus halving the time required for reliable detection of the target organisms using conventional methodology.

References

1.   D A A Mossel, *Food Technology in Australia*, 1978, **30**, 212.

2.   D A A Mossel, *Antonie van Leeuwenhoek*, 1982, **48**, 641.

3.   C L Baylis and S B Petitt, 'Coliforms and *E.coli* - Problem or Solution?', Royal Society of Chemistry, Cambridge, 1977.

4.   E F Drion and D A A Mossel, *Journal of Hygiene*, 1977, **78**, 301.

5.   G Suhren and W Heeschen, *Milchwissenschaft Milk Science International*, 1987, **42**, 619

6.   D L Cousins and F Marlatt, *Journal of Food Protection*, 1990, **53**, 568

7.   M Karwoski, *Food Reviews International*, 1996, **12**, 155

IMMUNOASSAYS FOR RAPID DETECTION OF FOODBORNE PATHOGENS AND TOXINS : A REVIEW

Purnendu C. Vasavada
Animal and Food Science Department
University of Wisconsin- River Falls
River Falls, WI 54022, USA.

1.0   INTRODUCTION

Imunoassays and other immunotechniques are the powerful and elegant techniques for rapid detection of foodborne pathogens and toxins. They also provide accurate and convenient means for detection of adulteration and verifying authenticity of foods[1]. The enzyme immunoassays (EIA) or enzyme-linked immunosorbent assays (ELISA) are the most common immunodiagnostic techniques used for rapid detection of foodborne pathogens including *Listeria monocytogenes, Salmonellae, E. coli* 0157:H7, *Campylobacter jejuni* and *Vibrio parahaemolyticus* and enterotoxins produced by *C. perfringens, Staphylococcus aureus* (SET), *E. Coli* 1057:H7 and *Bacillus cereus*[2-5]. The development of hybridoma technology to construct antibodies of predetermined specificity, new highly sensitive configurations and semi-or fully automated instrumentations for performing immunoassays have greatly increased the use of this technology for the rapid detection of micoorganisms and other analytes in food and environmental microbiology[6,7]. However, the complexities of food matrices and compatibility with conventional methods pose unique challenges that may limit universal direct application of these methods. This paper reviews developments, fundamental aspects, formats and applications of immunoassay technology in the context of microbiological safety of foods.

2.0   DEVELOPMENTS OF IMMUNOASSAYS

Historically, the first scientifically reported observation of an antigen-antibody reaction and generally accepted beginnings of Immunology as a science are traced to Jenner's research dealing with development of smallpox vaccine. The report of the interactions between a soluble antigen and a corresponding soluble antibody resulting in the formation of a precipitate by Kraus in 1897 and the establishment of immunoprecipitation in a gel medium by Bechhold in 1905 mark the probable beginning of immunoassays[8]. The first immunoassay, a radioimmunoassay (RIA) for insulin, was described by Yalow and Berson[9] in 1959. The RIA provided sensitivity, selectivity and a higher throughput than traditional assays and were widely accepted in the clinical field. The use of radioactive label was one of the factors that delayed acceptance of the RIA in the food industry, although the first use of RIA for the detection of specific proteins in food extracts was reported in 1970[8]. In 1971, Engvall and Perlmann[10] and Van Weemen and Schuurs[11] reported that radiolabel could be replaced by enzyme labels, thus marking the beginning of the enzyme immunoassays (EIA). The first report of EIA dealt with the detection of parasites, *Trichinnella spiralis*, in pigs for slaughter[8].

During the 1970's and early 80's there was a slow but steady increase in the publications describing food EIA's[8,13]. The type of analysis for which assays were developed included food constituents including proteins, caffeine, flavor components, antibiotics, toxins and hormones. By 1983, publications reporting EIA's and food analysis represented 45 to 56 % of total publications[13].

2.1   RECENT DEVELOPMENT

A computerized literature search of publications during 1969 - February 1999 in two of the most popular databases, viz. The Food Science and Technology Abstracts (FSTA) and the Biology and Agricultural Index (BAI), was conducted using key words "Immunoassays" or "ELISA" and various pathogens or toxins[12]. The FSTA search registered 59% more "hits" (2236 vs. 1323) than the BAI search for publications dealing with immunoassays. The BAI search revealed zero publications when the keywords "enterotoxin(s)" was used and registered only 73 hits when the keyword "toxin(s)" was used. The FSTA is apparently more appropriate data base for literature search dealing with food immunoassays. Earlier a survey by Clifford[13] had indicated that majority of food immunoassays are found most conveniently in the FSTA. The FSTA search indicated that there was a strong increase in the applications of immunological methods concerning *emerging* pathogens such as *E. coli* 0157:H7, *Listeria monocytogenes* and *Campylobacter jejuni* during the 90's. The review of current literature indicated: 1) A high proportions of immunoassays (ELISAs), 2) A smaller proportion of immunoprecipitation, 3) Increased use of monoctonal antibodies, and 4) Increased use of immunoassay for food allergens, hormones, antibiotic residues[12].

3.0   IMMUNOASSAY FUNDAMENTALS

Immunoassay techniques are methods that exploit the specific interaction of antigen (Ag) and corresponding antibody (Ab) to form an antigen-antibody (Ag-Ab) complexes that are readily detectable using enzyme or radio labels. Enzyme immunoassays allow detection of trace amount of analytes with little or no requirement for purification and/or concentration of the sample and are method of choice in food and other non-clinical fields. Radio-immunoassays are seldom used in food analysis because of the short half-life of many radioactive materials, special handling and disposal requirements and the undesirability of the presence of radioisosopes in food processing environments.

3.1   IMMUNOLOGICAL TERMS

Basic understanding of immunoassay techniques requires familiarity with certain immunological terms which are defined and explained below. Detailed discussion of immunological terms can be found elsewhere[8,14].

Antibody (Ab).

Antibodies are proteins produced by immune cells of the body in response to administration of an antigen. Antibodies react with antigen or sometimes, with substance of similar structure and form Ag-Ab complex. Advances in large scale production of specific antibodies have been significant in commercializing immunoassay techniques such as ELISA. Two types of antibodies can be produced: **Polyclonal** and **Monoclonal**. Antibodies produced by conventional technology through the immunization of animals are polyclonal antibodies. Individual lymphocytes in the animal do not always produce exactly the same antibody molecule in response to a given antigen. Thus the final mixture of antibody can be heterogeneous in specificity and may lead to false-positive or false-negative results. A polyclonal antibody preparation may contain up to 10,000 different types of IgG and most likely produce binding over a wide range of conditions. They may provide a more robust system. Monodonal antibodies are produced by in vitro cultures of cells derived from a single isolated lymphocyte. These antibodies are chemically identical, exhibit a high degree of specificity and are often preferred for use in enzyme immunoassays.

Antigen (Ag).

Antigens are substances which are capable of inducing the formation of antibodies and of reacting specifically in some detectable manner with the antibodies so induced. Immunogens are the substance that induce production of antibodies and the ability to induce formation of antibodies is referred to as Immunogenicity. Antigens and immunogens are different from **haptens** which are small chemical moieties that can react specifically with certain antibodies. They are different from antigens in that they cannot induce antibody production.

Epitope

An epitope is the region of an antigen that reacts with the antibody. It is not an intrinsic property of any particular structure and is defined by reference to the binding site of an antibody. Since antibodies can recognize relatively small regions of antigens, occasionally they can find similar epitopes on other molecule thereby causing cross-reaction. Similarly, small changes in the epitope structure can prevent antigen recognition.

Affinity and Avidity

Affinity is a measure of the strength of the binding of an epitope to an antibody and, as such, is indicative of the amount Ag-Ab complex formed during the immunological reaction. High-affinity antibodies bind larger amount of antigen in a shorter period of time than low-affinity antibodies. High-affinity antibodies have higher capacity for binding with antigens, form stable Ag-Ab complex in a few minutes and are preferable in immunoassay techniques. Avidity is a measure of the overall stability of the Ag-Ab complex and important factor in determining the ultimate success of immuno chemical technique.

3.2   IMMUNOASSAY FORMATS AND CONFIGURATIONS

Immunoassays can be classified as homogeneous and heterogeneous. In a homogeneous assay, the antigen detection actually modifies the enzyme detection of the reaction. The specific antibody targets are haptens coupled with enzymes or substrate, coenzyme, or inhibitors of enzyme. There is no need to separate the bound and free antibodies since the assay is performed in liquid phase. Examples of homogeneous assays are agglutination reaction and particle-enhanced turbudimetric inhibition immunoassay. Homogeneous assays require shorter incubation times, can be quantitative and are amenable to automation. Heterogeneous assays require separation of bound and free reactants and therefore, require a solid phase to immobilize one component of the reaction to facilitate the separation of unbound reagents. Most ELISA's used in food analysis are examples of heterogeneous immunoassay. The components of a typical heterogeneous immunoassay include: analyte capture, reporter addition, washing steps, and reporter detection. In a typical sandwich ELISA, a capture antibody is immobilized on a solid support, sample or enriched sample is added followed by incubation to facilitate Ag-Ab complex. The unbound Ag is removed by washing and a reporter antibody conjugate is added. After a brief incubation, the unbound reporter conjugate is removed by a wash and a reporting substance is added. The amount of reporter signal is proportioned to the amount of analyte in the sample.

Popular configurations and assay formats[15] are used in immunoassays for detection foodborne pathogens and toxins include: 1) Latex Agglutination (LA) and Reverse Passive Latex Agglutination (RPLA), 2) Immunodiffusion, 3) ELISA, 4) Immunoprecipation, and 5) Immunomagnetic separation (IMS). (Table 1).

The latex agglutination is the simplest assay in which Ab coated color latex beads are used to agglutinate specific antigens to form visible clumping or precipitate. Although simple and convenient, it is not a very sensitive assay and is mostly used for confirming the identity of pure cultures. In RPLA antibodies are attached to latex beads and used to detect bacterial toxins.

Immunodiffusion is another simple technique in which an antigen (motile pathogen) is detected by a line of precipitation formed by the reaction of antigen with diffusing antibody. This technique would not detect nonmotile pathogens.

The ELISA is probably the most popular immunoassay technique used for detection of foodborne pathogens. The "sandwich" assay using enzyme label, colorimetric detection, and microtiter plate as a solid matrix is the most common assay configuration. The detection sensitivity of ELISA is about 10^5 CFU/ml for whole bacterial cells and a few ng/ml for toxin or protein analytes. Many operations in performing ELISA including pipetting reagents and washing steps have been mechanized and fully automated. ELISA systems have been developed to facilitate sample throughput and minimize false-positive results.

Immunoprecipitation assays are essentially "sandwich" antibody assays in which detection antibodies labeled with gold or colored latex beads are used instead of enzyme conjugates. The assays are similar to home pregnancy tests and are extremely simple, rapid and user-friendly.

The immunomagnetic separation (IMS) uses magnetic beads coated with specific antibodies to capture specific pathogens from pre-enrichment samples, thus eliminating the need for selective enrichment or other enrichment steps. The IMS can be coupled with ELISA or other analytical methods such as PCR test. Assays based on selective concentration or capture of bacterial pathogens using IMS/ELISA have been developed for foodborne pathogens viz. *Listeria, E. coli* 0157:H7 *and Salmonella*.

Recent technological advances in instrumentations, consumables and reagents has lead to development of partially or fully automated systems for conducting immunoassays. These instruments can process multiple samples, computerize data handling and analysis, reduce labor costs and improve accuracy of assay[4,5,15]. The Vidas (biomerieux), the OPUS (Tecra), and the EIA Foss (Foss Electric) are examples of commercially available automated immunoassay systems. These systems seem to provide reliable detection of *Salmonella, Listeria, E. coli* 0157 and other pathogens and toxins and are becoming increasingly popular in the food industry[4]. Detailed discussion on various formats, configurations and components of immunoassay techniques may be found elsewhere[5-7, 13, 16, 17].

4.0   LIMITATIONS AND CHALLENGES

Immunoassay technology offers several advantages over the conventional cultural methods for detection of foodborne pathogens and toxins including speed, convenience, user friendliness and relatively high degree of specificity and reliability. They also have several limitations, e.g. they cannot reliably detect $<10^4$ - 10^5 microbial cells/ml. Since current regulations require detection of ≤ 1 cells in 25g food sample, the sensitivity of assay can be important challenge facing some immunoassay methods. Pre- and selective enrichment protocols designed to improve sensitivity and selectivity of the assay require extra steps and 12-12 h to the time of detection.

Other limitations and challenges of immunoassay technique are non-specific reactions and cross reactivity leading to false positive and false negative results. False-positive results occur when antigen other than that of the target species of toxin react with specific antibody used in the assay. The purity and specificity of the antibody used and non-specific binding of antibody (Ab-enzyme conjugate) to sites on solid support may be critical in obtaining accurate results. In case of a toxin, the antibody may still react with the toxin even if it has lost its biological activity, thereby giving a false-positive result. Some foods contain cross-reactive substances which bind to the antibodies used in the assay leading to a false-positive result. False-positive reactions can be recognized by analyzing the reaction kinetics and by specific blocking of the IgG by synthetic epitopes or by anti-idiotype antibodies. The proportion of false-positive reactions by an assay must be kept to a minimum ($< 5\%$).

Table 1. Partial-listing of commercially-available, antibody-based assays for the detection of foodborne pathogens and toxins[11]

Organisms/toxin	Trade Name	Assay Format[a]	Manufacturer
Bacillus cereus diarrhoeal enterotoxin	TECRA	ELISA	TECRA, Roseville, Australia
	BCET-RPLA	RPLA	Unipath (Oxoid) Ogdensburg, NY
Campylobacter	Campyslide Merutec-campy MicroScreen	LA LA LA	Becton Dickinson Cockeysville, MD Meridian Diagnostic Cincinnati, OH Mercia Diagnostics Shalford, UK
Clostridium botulinum toxin	ELCA	ELISA	Elcatech Winston-Salem, NC
C. perfringens enterotoxin	PET-RPLA	RPLA	Unipath (Oxoid)
Escherichia coli EHEC[c]	RIM	LA	REMEL, Lenexa, KS
E. coli 0157/0157:H7	E. coli 0157 Prolex Ecolex 0157 Petrifilm HEC EZ COLI Dynabeads HEC 0157 EHEC-TEK Assurance TECRA E. coli 0157 VIP[d] Reveal NOW	LA LA LA blot tube-EIA IMS ELISA ELISA ELISA ELISA ELISA IPPT IPPT IPPT	Unipath PRO-LAB, Austin, TX Orion Diagnostica Somerset, NJ 3M, St. Paul, MN Difco, Detroit, MI Dynal, Lake Success, NY 3M, Canada Organon-Teknika Durham, NC BioControl, Bothel, WA TECRA LMD, Carlsbad, CA BioControl, Neogen, Lansing, MI Binax, Portland, ME
STX-1 & 2	VEROTEST Premier EHEC Verotox-F	ELISA ELISA RPLA	Microcarb, Gaithersburg, MD Meridian, Cincinnati, OH Denka Seiken, Tokyo, Japan

Organisms/toxin	Trade Name	Assay Format[a]	Manufacturer
Salmonella	Bactigen	LA	Wampole Labs Cranbury, NJ
	Spectate	LA	Rhone-Poulenc Glasgow, UK
	Microscreen	LA	Mercia Diagnostic Surrey, UK
	Wellcolex	LA	Laboratoire Wellcome Paris, France
	Serobact	LA	REMEL, Lenexa, KS
	RAPIDTEST	LA	Unipath (Oxoid)
	1-2 Test		BioControl
	Dynabeads	IMS	Dynal
	Screen	IMS	VICAM
	CHECKPOINT	blot	KPL, Gaithersburg, MD
	Salmonella-TEK	ELISA	Organon Teknika
	TECRA[c]	ELISA	TECRA
	EQUATE	ELISA	Binax
	BacTrace	ELISA	KPL
	Assurance	ELISA	BioControl
	Salmonella	ELISA	GEM, Hamden, CT
	LOCATE	ELISA	Rhone-Poulenc
	UNIQUE	capture-ELISA	TECRA
	PATH-STIK	IPPT	Lumac, The Netherlands
	Reveal	IPPT	Neogen
	Clearview	IPPT	Unipath (Oxoid)
Staphylococcus aureus	Staphylo-slide	LA	Becton Dickinson Cockeysville, MD
	Aureus Test	LA	Trisum, Taipei, Taiwan
	TECRA	ELISA	TECRA
Staph enterotoxins (SET)	SET-RPLA	RPLA	Unipath (Oxoid)
	TECRA[c]	ELISA	TECRA
	SET-EIA	ELISA	Toxin Technology, Madison, WI
	TRANSIA	ELISA	Transia, Lyon, France
	RIDASCREEN	ELISA	R-Biopharm, Germany
Vibrio cholera	choleraSMART	IPPT	New Horizon Columbia, MD
	bengalSMART	IPPT	New Horizon
enterotoxin	VET-RPLA	RPLA	Unipath (Oxoid)

[a] Abbreviations ELISA, enzyme linked immunosorbent assay; LA, latex agglutination; RPLA, reverse passive LA; IPPT, immunoprecipitation; IMS, antibody-magnetic beads.
[b] Enterohemorrhagic *E. coli.*
[c] Also available as BioPro (International Bioproducts, Redmond, WA).

False-negative results occur if the target organism does not grow to a minimum populations of 10^5 - 10^6 cfu/ml during enrichment culture. A false negative reaction may also be due to the failure of production or expression of specific antigenic molecule, e.g. flagella production.

Food analysis using immunoassays pose other important challenges. Immunological structures may be modified during performance of an immunoassay. Also, some treatments (heat, exposure to acid or alkali, vigorous agitation or extrusion, etc.) may lead to irreversible changes in some proteins, thereby potentially changing reactivity to specific antibodies e.g. In the detection of *Staphylococcus aureus* enterotoxins in heat processed foods the ELISA may yield a negative test if heating inactivated the antigenic site on the toxin but biological activity remain unaffected. To obtain accurate test, samples must be renatured before analyzing by ELISA.

5.0   FUTURE TRENDS

Immunoassays and other immunote techniques to detect foodborne pathogens and toxins were scarcely known twenty years ago. The Delphi forecast[18] conducted in 1981 to predict trends and developments in food microbiology failed to forecast the potential of immunological techniques on food diagnostics.[19]

The rapid developments in immunoassay techniques over the past two decades have been directed towards increasing specificity, sensitivity and reliability as well as automation of the various techniques. New applications of immunoassay technique in the food industry viz., detection of aduteration, authentication of food ingredients and raw material and detection of food allergens are being developed. Automation, miniaturization and computers applications will provide new and improved immunoassay techniques combining rapid reliable results with high sample throughput, hopefully without sacrificing simplicity and elegance of the immunoassay methods. Another exciting development is the potential application of antibodies in biosensors. While the use of biosensors in food analysis is still in early stages of development, immunosensors have been tested in which an antibody is coupled to piezoeleitric crystals were found to detect 10^6 E. coli cells/ml.[20] Although actual testing of pathogens or toxins in contaminated foods is still not feasible, the potential development of immunosensor that can be used in direct, online or at-line testing thus providing real time food safety assessment is an exciting possibility. Ultimately improvements in immunoassays could lead to the development of simple, reliable, economic and user-friendly 'dip-stick' assays which require no interpretation and minimum skill to use. Of course, the regulatory implications of these new methods in the context of 'zero tolerance' for pathogens and impact on the food industry resulting from more sensible detection methods remain the subject of dialog between the food industry, regulatory agencies and the consumer.

6.0 ACKNOWLEDGMENT

Contribution of the College of Agriculture, Food and Environmental Science, University of Wisconsin-River Falls and the University Wisconsin Cooperative Extension. The mention of specific companies, methods or instruments is for general reference only and does not necessarily indicate verification, endorsement or approval by the author or the UW-RF and UW-Extension.

REFERENCES

1. IFT Symposium, *Food Technol*, 1995, **49(2)**,101.
2. P. Feng, *Molecular Biotechnology*, 1997, **7**,267.
3. B. Swaminathan and P. Feng, *Annu. Rev. Microbiol.*, 1994, **48**,40.
4. P. C. Vasavada, *Food Testing and Analysis*, **3(2)**,18.
5. J. M. Cox and G. H. Fleet, "Foodborne Microorganisms of Public Health Significance", A. D. Hocking, ed., AIFST, NSW Branch, Food Microbiology Group, Sydney, Australia, 1997, f ifth edition, Chapter 2, p. 85.
6. G. M. Wyatt, H. A. Lee and M. R. A. Morgan, "Immunoassays for Food Poisoning Bacteria and Bacterial Toxins", Chapman and Hall, London, 1992.
7. J. H. Rittenburg, 1990, "Development and Application of Immunoassay for Food Analysis", Elsevier Applied Science, London & NY. 1990.
8. C. J. Smith, "Development and Application of Immunoassay for Food Analysis", J. H. Rittenburg, ed., Elsevier Applied Science, London, 1990, Chapter 1, p. 3.
9. R. S. Yalow and S. A. Berson, *Nature*, 1959, **184**,1643.
10. E. Engvall and P. Perlmann, *Immunochemistry*, 1971, **8**,87.
11. B. K. VanWeemen and A. H. W. M. Schurrs, *FEBS Letters*, **13**,232.
12. P. C. Vasavada, and T. Fruth, 1999, unpublished.
13. M. N. Clifford, "Immunoassays in Food Analysis", B. A. Morris and M. N. Clifford, ed., Elsevier Applied Science Publishers, London, Chapter 1, p. 3.
14. B. A. Morris, "Immunoassay in Food Analysis", B. A. Morris and M. N. Clifford, ed., Elsevier Applied Science Publishers, London, p. xv.
15. P. Feng, "Bacteriological Analytical Manual", AOAC International, Gaithersburg, MD, 8th ed., 1998 Rev. A.
16. M. R. A. Morgan, C. J. Smith and P. A. Williams, "Food Safety and Quality Assurance: Applications of Immunoassay Systems", Elsevier Applied Science, London, 1992.
17. W. M. Barbour and G. Tice, "Food Microbiology Fundamentals and Frontiers", M. P. Doyle, L. R. Beuchat and T. J. Montville, ed., ASM Press, Washington, DC, 1997, Chapter 32, p. 710.
18. B. Jarvis, "Rapid Methods and Automation in Mcirobiology and Immunology", K. O. Habermehl, ed., Springer-Verlag, Berlin, p. 593.
19. P. C. Vasavada, *J. Rapids Methods and Automation in Microbiol.*, 1993, **2**, 1.
20. S. Oh, *Trends Food Sci Technol.* 1993, **4**,98.

USE OF BACTERIOPHAGE FOR RAPID DETECTION OF MICRO-ORGANISMS

Richard J. Mole,

Biotec Laboratories Ltd.,
32 Anson Road,
Martlesham Heath,
Ipswich, IP5 3RG.

1. INTRODUCTION

Bacteriophage are well recognised tools for the typing of bacteria (phage typing) and in molecular biology (for instance cloning in bacteriophage λ). However bacteriophage also posses several attributes which make them suited for the detection of bacteria. These principally are their specificity for groups of bacteria and their rapid growth and biochemistry, which permits their use as reporters of bacterial presence and viability.

Three uses of bacteriophage will be discussed (Figure 1). Recombinant bacteriophage, such as *lux*-bacteriophage, infect a target bacterium and the recombinant gene product produces a distinguishable signal reporting the presence of the target bacterium. An alternative method (Phage Amplification) is to use the bacteriophage themselves a reporter by allowing the propagation of progeny phage which are then detected, such as a zone of clearing in a lawn of helper cells. Lastly, bacteriophage encode enzymes that are responsible for the lysis of host cells, which can be incorporated into the ATP-bioluminescence assay to add specificity to current hygiene monitoring systems (Lysin release ATP-bioluminescence).

2. BACTERIOPHAGE ATTRIBUTES EXPLOITED BACTERIAL DETECTION

A well recognised property of bacteriophage is their specificity to infect defined ranges of bacteria, normally within the same genus. This attribute is used within phage detection technology to give specificity to the detection of certain types of organisms. For instance, if *Campylobacter* were to be detected, a *Campylobacter* specific bacteriophage would be used. Dead cells are not metabolically active, hence, do not support bacteriophage replication, and are not detected by bacteriophage-based methods. Bacteriophage replication within a cells is rapid, compared to cell growth and yields typically 20-100 progeny particles in the same time period as the doubling time of it's host. This can be harnessed to give a much higher signal in a shorter time frame than cell growth. To facilitate this rapid replication, bacteriophage contain strong gene promoters, which if linked to a reporter gene, permits rapid and substantial expression of a signal.

Figure 1 Perceieved benefits and difficulties with bacteriophage detection techniques.

	Advantages	Disadvantages
Recombinant bacteriophage (*lux*-phage)	• Rapid (1-2 hours) • Specific • Sensitive • Quantitative • Only detects viable cells	• Genetically modified phage • Equipment orientated • Host range of phage • Food matrix may interfere with detection
Phage Amplification	• Rapid (4-10 hours) • Specific • Sensitive • Quantitative • Simple • Only detects viable cells	• Host range of phage
Lysin release ATP-bioluminesence	• Extremely rapid (10 min) • Specific • Quantitative • Compatible with current ATP assay • Only detects viable cells	• Equipement orientated • Low sensitivity

3. RECOMBINANT PHAGE TECHNOLOGY

Several reporter genes have been inserted into bacteriophage genomes, including:
1. *luxAB* (2kbp) that encodes for the bacterial luciferase gene, and confers a light emitting phenotype to infected cells, which can be detected by either light sensitive cameras, luminometers or photographic film.
2. *ina* (3.6kbp) that encodes for the bacterial ice nucleation protein, which acts as a template for ice nucleation and can be detected as colour change in a dye as water freezes.

A recombinant *Listeria* bacteriophage (A511::*luxAB*) containing the *luxAB* gene has been used for the detection of a number of Listeria strains (1). A detection limit of 100 cell/ml was observed, below which background light interfered with the signal. Detection was semi-quantitative, with increasing numbers of cells giving more light.

Improved detection was seen after a 20hr enrichment step with detection of 0.1 cfu/g, 1 cfu/g and 10 cfu/g observed with cabbage, milk and camembert cheese, respectively (2). *lux*-bacteriophage have also been successfully incorporated into most probable number methods for bacterial enumeration for *Listeria, Salmonella* and *E. coli*.

4. PHAGE AMPLIFICATION

The principles of the Phage Amplification technique are as follows (Figure 2):

1. Infection stage- bacteriophage attach and infect target bacteria then start to replicate within the cell.
2. Virucide stage - exogenous phage are destroyed by a potent virucide (JSD solution) which acts selectively against the bacteriophage, whist remaining not detrimental to the cells. Hence, only the bacteriophage within the target cells survive this process.
3. Neutralisation stage - virucide is neutralised so that when progeny phage are release they are not inactivated.
4. Amplification (Helper) stage - helper cells (typically non-pathogenic, fast growing, phage susceptible strains) are added. Progeny bacteriophage released from infected cells replicate in the helper cells, leading to amplification of the phage signal. If the sample is placed in an agar overlay, the helper cells will form a lawn of growth and released bacteriophage will cause zones of clearing (plaques), representative of the number of cells in the original sample.

The technique has been shown to work successfully for *Campylobacter, Salmonella, Pseudomonas, Listeria, Staplococcus, E. coli* and *Mycobacteria*. Detection sensitivities of 120 cfu/ml, 40 cfu/ml and 600 cfu/ml has been demonstrated for *Campylobacter, Salmonella* and *Pseudomonas*. These figures could be improved by analysis of larger samples (currently 10-100μl).

Detection of *Campylobacter* cells in the presence of large numbers of competitive bacteria and in food substrates (chicken homogenate) has been demonstrated.

5. BACTERIOPHAGE LYSIN- BASED DETECTION

Bacteriophage release from host cells towards the end of their replicative cycle is often mediated by two phage encoded proteins. A holin protein forms a lesion in the cell membrane which permits the passage of a lysin enzyme through the membrane to the cell wall. The lysin attacks on the peptidoglycan layer, resulting in the loss of cell wall structural integrity, causing lysis of the cell and dispersal of the cellular contents (including ATP) and the bacteriophage. A *Listeria* bacteriophage lysin has been shown to have a specificity for the cell walls of the *Listeria* genus. This discovery has been exploited to give specificity to the otherwise non-specific ATP-bioluminescence detection method by replacement of the general lysing reagent with the lysin enzyme. Detection of axenic and mixed cultures of *L. monocyotgenes* and *L. ivanoii* has been demonstrated (3).

Figure 2. The principle of Phage Amplification technique for the detection of Campylobacter cells

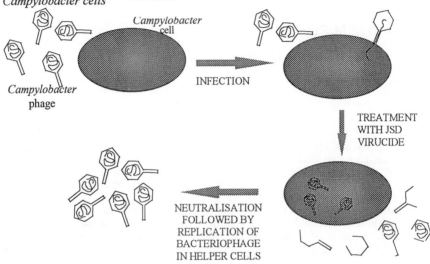

1. M. J. Loessner, C. E. D. Rees, G. S. A. B. Stewart and S. Scherer. Appl. Env. Micro., 1996, **62**, 1133.

2. M. J. Loessner, M. Rudolf and S. Scherer. Appl. Env. Micro., 1997, **63**, 2961.

3. G. S. A. B. Stewart, M. J. Loessner and S. Scherer. ASM News, 1996, **62**, 297.

RAPID METHODS FOR ENUMERATING THE HETEROTROPHIC BACTERIA IN BOTTLED NATURAL MINERAL WATER

M. Fitzgerald, M. Kerr and J. J. Sheridan

Teagasc, The National Food Centre,
Dunsinea, Castleknock,
Dublin 15,
Ireland.

1 INTRODUCTION

Over the past two decades, the UK bottled water market has developed from annual bottled water sales of 30 million litres in 1980 to 800 million litres in 1995.[1] This increase in demand may be principally driven by changes in fashion towards the conspicuous consumption of "designer water", but it may also reflect increased consumer concerns in relation to the perceptions of the safety of mains supply water. [2]

Due to this increased demand it is important to be able to assure the consumer of the quality and safety of bottled mineral waters. European regulations were instituted to control the bottling and marketing of natural mineral water (Directives 80/777/EEC and 96/70/EC).[3,4] Detection of pathogens and quantification of bacterial numbers is important in the mineral water industry in order to comply with these EU regulations. At present the conventional method for determining the heterotrophic plate count (HPC) takes up to three days. The delay in awaiting microbiological results is time consuming, delaying both the identification of hygiene problems and the release of products to the market place. Therefore, the development of methods that are capable of reducing the time required to obtain a bacterial count, ideally from days to minutes are urgently required by the industry.

Direct microscopy has long been established as a simple rapid and inexpensive method for the enumeration of microorganisms. The direct epifluorescent technique (DEFT) using acridine orange (AO) was developed to rapidly assess the microbiological quality of raw milk.[5] This technique has also been applied to products such as beef,[6] pre-filtered food suspensions, beverages and water.[7] The notable disadvantage of this technique is the inability of AO to distinguish between metabolically active and dead cells. However, a direct technique for enumerating metabolically active bacteria involving the use of AO has been developed.[8] The Kogure method involves pre-incubation of water samples with yeast extract and nalidixic acid. This technique is based on the specific inhibition of DNA synthesis by nalidixic acid. Active cells unable to divide due to the lack of DNA synthesis will continue to grow and elongate in the presence of the yeast extract. These elongated/viable cells can be directly enumerated microscopically following staining with AO.

A number of alternative techniques which allow direct determination of actively metabolising bacteria from filtered samples have been developed. Both 2-(p-iodophenyl)-3-(p-nitrophenyl)-5-phenyl tetrazolium chloride (INT) and 5-cyano-2, 3-di-4-tolyl tetrazolium chloride (CTC) have been used to assay and enumerate respiring and metabolically active bacteria in aquatic environmental samples.[9,10] These techniques are based on the principle that the tetrazolium dyes are reduced by electron transport activity to give insoluble crystals of formazan, which accumulate in metabolically active bacteria, and can be viewed by microscopy. The LIVE/DEAD BacLight bacteria viability kit (BL) is also capable of detecting viable cells, this method has been successfully applied as a rapid direct viable count method for the determination of total viable counts on processed meats.[11] The kit contains a mixture of two fluorescent nucleic acid stains, which stain live bacteria with intact membranes green and dead cells with disrupted membranes orange. The ChemScan system is another analytical method which has been used for the rapid detection and enumeration of viable microorganisms in water samples.[12] This method is based on the direct fluorescent labelling of microbial cells on membrane filters and subsequent detection by a laser scanner. The viable microorganisms are labelled using a non-fluorescent fluorescein derivative called Chem-Chrome (CC), which is cleaved by esterase activity within the cell to give the fluorescent product, fluorescein. In addition to esterase activity, viability staining by fluorescein derivatives is based on the principle that only viable cells, which have an intact membrane, are able to retain and accumulate the fluorescent probe.[12]

The objective of this study was to investigate the use of fluorochromes in conjunction with epifluorescent microscopy to rapidly enumerate the autochthonous flora present in mineral water and to compare these direct counts with those obtained by the HPC, to determine if any of these rapid methods could be used to predict the HPC at 22 °C and/or 37 °C. Four fluorochromes were used, AO, *BL* (Molecular Probes, Leiden, The Netherlands), CC (Chemunex, Paris, France) and CTC (Polysciences Europe GmbH, Eppelheim, Germany).

2 MATERIALS AND METHODS

2.1 Water Samples

One batch of a European brand of natural non-carbonated mineral water (MW) bottled in polyethylene terephthalate (PET) bottles was obtained directly from the mineral water company within 24 h of bottling. On delivery to the laboratory the water was stored at 2 °C for up to 52 days. Additionally, a random selection of different batches of MW samples were obtained from retail outlets and the mineral water company.

2.2 Analysis of Water Samples for the Generation of Prediction Equations

Samples of MW were removed from the batch of water stored at 2 °C every three or four days over a period of 52 days following bottling. The numbers of bacteria present in these samples were enumerated on each sampling occasion by the rapid direct counting techniques and by the HPC.

2.2.1 Rapid Direct Counts. MW was vacuum filtered through black polycarbonate membranes (0.2 μm pore size, 25 mm diameter; Poretics, Osmonics, Ca., USA). Initially, when bacterial numbers were low following bottling, one to two litres of water were analysed, as the numbers increased, smaller volumes were analysed. Each membrane was stained with one of the following dyes.

AO: The membrane was overlaid with 1 ml 0.025% (w/v) AO (Gurr) for 1 min. Excess stain was rinsed from the membrane using 50 μl industrial methylated spirits. The membrane was air dried and mounted in a non-fluorescent immersion oil on a glass slide beneath a coverslip.

BL: The membrane was stained with *BL* using the procedure previously described.[11]

CC: The CC working solution was prepared by adding 100 μl Chem-Chrome V2 viability substrate to 10 ml of filter-sterilised Chemsol B4 marking solution. A 47 mm filter pad (Whatman) was placed in a 50 mm petri dish (Sterilin) and the pad was impregnated with 2 ml of CC working solution. A black membrane through which mineral water had been filtered was placed on the moistened pad and incubated at 37 °C for 30 min. The membrane was air dried and mounted as described above.

CTC: To a 47 mm filter pad placed in a 50 mm petri dish, 2 ml of 5 mM CTC solution prepared in 0.85% (w/v) filter sterilised sodium chloride was added. A black membrane through which mineral water had been filtered was placed on the moistened pad and incubated at 37 °C for 1.5 h. The membrane was air dried and mounted as described above. The stained membranes were examined under the 100 X oil immersion objective of a Nikon epifluorescent microscope with a 100 W mercury vapour light source. On the membranes yielding between 5 and 20 cells per field of vision, 20 fields were counted. The number of bacteria per millilitre of sample was then calculated by multiplying the total number of bacteria in 20 fields by a microscope working factor and dividing this by the volume of MW filtered.

2.2.2 Heterotrophic Plate Count (HPC). The HPCs of the bottled MW were enumerated on the non-selective agar, R2A. Initially when bacterial numbers were low following bottling the HPC was determined using membrane filtration, 1, 10 100 and 500 ml volumes were analysed on each sampling occasion. As the numbers of bacteria increased in the bottles the HPC was determined by the spread plate technique. Two separate sets of plates were prepared in duplicate, one set was incubated at 37 °C for 1 day, while the other set was incubated at 22 °C for 3 days.

2.3 Analysis of Water Samples for the Validation of the Prediction Equations

In order to validate the prediction equations derived from the *BL* and CTC counts, a random selection of bottled non-carbonated mineral water samples (n=34) were examined. The *BL* and CTC rapid direct counts and the HPC at 22 and 37 °C for all these samples were determined as described above.

2.4 Statistical Analysis

The relationships between each of the rapid direst counts (AO, *BL*, CC and CTC) and the HPCs at 37 °C and at 22 °C were evaluated by regression analysis. The relationship between the actual plate counts of the random selection of MW samples were compared with the plate counts predicted by equations derived previously from the *BL* and CTC counts, using regression analysis.

3 RESULTS

The best correlation was obtained between the HPC at 22 °C and the *BL* count ($r^2 = 0.93$, rsd = ±0.42) (Figure 1). A good correlation was also obtained between the HPC at 22 °C and the CTC count ($r^2 = 0.85$, rsd = ±0.62). Very poor correlations were obtained between the counts obtained with any of the four rapid direct count methods and the HPC at 37 °C ($r^2 < 0.2$). Since the prediction equations relating *BL* and CTC counts to the HPC at 22 °C gave the best correlations, these equations were validated by analysing a random selection of mineral water samples. A linear relationship was observed between the actual plate count at 22 °C and the counts predicted by both the *BL* count ($r^2 = 0.86$, rsd = ±0.5) (Figure 2) and the CTC count ($r^2 = 0.82$, rsd = ±0.62), indicating that the equations derived from the membrane counts using *BL* or CTC correlate well with the HPC at 22 °C on R2A agar.

4 CONCLUSIONS

The results indicate that both the *BL* and the CTC direct counting techniques have potential in rapidly predicting the HPC at 22 °C for bottled non-carbonated mineral water. None of the four rapid direct counting techniques were suitable in predicting the HPC at 37 °C. The *BL* method is currently being tested by a leading European bottled mineral water company.

Either the *BL* or CTC technique should be of benefit to the mineral water industry because of their speed and relative ease of use. A result is available by the *BL* method within 30 min and by the CTC method within 2 h. The current plate count technique takes 3 days to complete. These techniques would be of far more value to the mineral water industry if the enumeration of the stained bacteria could be automated, thereby, increasing the numbers samples that can be analysed per day and eliminating the chances of operator fatigue due to prolonged use of the microscope. Currently, there is no suitable or cost-effective technology available capable of distinguishing between the two

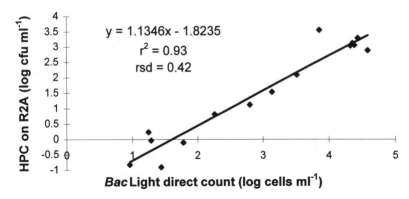

Figure 1 *Relationship between the heterotrophic plate counts of mineral water flora on R2A Agar at 22 °C for 3 days and the direct BacLight count*

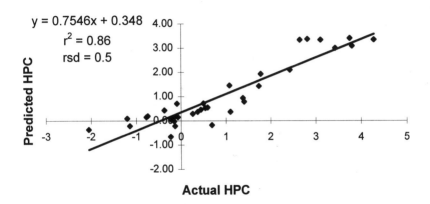

Figure 2 *Relationship between the actual HPC and the HPC as predicted by the BacLight using the equation previously derived (Figure 1).*

colour fluoresence of the *BL* stain or of detecting CTC stained cells. If the detection and the enumeration of *BL* and CTC stained cells could be automated it would be interesting to determine if the relationships between the rapid direct counts and the HPC could be improved. However, it may be difficult to substantially improve on these relationships as there will always be a discrepancy between the counts obtained by the rapid methods and the HPC due to the fact that the HPC fails to detect viable but non-culturable bacteria that may be present in aquatic environments such as mineral water.

References
1. Zenith International Ltd., "The Zenith Guide to the UK Bottled Water Market - December 1996", Zenith International, Bath, UK, 1996.
2. P. R. Hunter, *Microbiol. Eur.*, 1994, **2**, 8.
3. Anonymous, *Official J. Eur. Communities*, 1980, **L229**, 1.
4. Anonymous, *Official J. Eur. Communities*, 1996, **L299**, 26.

5. G. L. Pettipher, R. Mansell, C. H. McKinnon and C. M. Cousins, *Appl. Environ. Microbiol.*, 1980, **39**, 423.
6. I. Walls, J. J. Sheridan and P.N. Levett, *Ir. J. Food Sci. Tech.*, 1989, **13**, 23.
7. G. L. Pettipher, "The Direct Epifluorescent Filter Technique for the Rapid Enumeration of Microorganisms. Innovation in Microbiology Series 1", Letchworth Research Studies Press, 1983.
8. K. Kogure, U. Simidu and N. Taga, *Can. J. Microbiol.*, 1979, **25**, 415.
9. R. Zimmermann, R. Iturriaga and J. Becker-Birck, *Appl. Environ. Microbiol.*, 1978, **36**, 926.
10. G. G Rodriguez, D. Phipps, K. Ishiguro and H. F Ridgway, *Appl. Environ. Microbiol.*, 1992, **58**, 1801.
11. G. Duffy and J. J. Sheridan, *J. Microbiol. Meth.*, 1998, **31**, 167.
12. G. Wallner, D. Tillmann, K. Haberer, P. Cornet and J.-L. Drocourt, *Eur. J. Parenteral Sci.*, 1997, **2**, 123.

DEVELOPMENT OF METHODS TO DETECT FOODBORNE VIRUSES

A.S. Kurdziel, N. Wilkinson, S.H. Gordon and **N. Cook**

Food Microbiology Group
Central Science Laboratory,
Sand Hutton,
York YO41 1LZ

1   INTRODUCTION

Several types of food, for example shellfish, soft fruit and salad vegetables, have been involved in transmission of human enteric viruses. These foods have in common that they are generally eaten raw, or only lightly cooked. There has been some published methodology designed to detect viruses in these foodstuffs (mainly shellfish)[1,2]. Initial methods were based upon detection of viruses by their infection of cultured mammalian cells. However, not all virus types are readily detected by this means: small round structured viruses, the most significant foodborne viral agent of gastroenteritis in the UK[3], do not infect currently available cell lines. Recent detection protocols have been based upon nucleic acid amplification by the polymerase chain reaction (PCR) and its variants, and extraction techniques have been devised to deliver target viral sequences to the small volume tubes in which amplification reactions proceed. Most of these methods have however been so complex and time consuming that they are unlikely to be practically applied; furthermore, they can often also concentrate substances inhibitory to the amplification reaction, resulting in inconsistent results and low sensitivity.

We have been developing rapid methods to extract viruses from soft fruit and salad vegetables, using protocols based on differential flocculation and centrifugation. Poliovirus has been used as a model to monitor the efficiency of the extraction procedures, by quantal cell culture. The procedures are currently being refined to make them suitable for delivery of viral particles to nucleic acid amplification systems.

2   EXTRACTION PROTOCOLS

2.1 Salad Vegetables

The following protocol has been developed, for use with lettuce, green onions and white cabbage. The food sample size is determined by the average portion for consumption[4], i.e. 30 g lettuce, 10 g green onion, and 90 g white cabbage.

1. Place sample in filtered stomacher bag
2. Add 90 ml alkaline buffer
3. Stomach or pulsify 1 min

4. Decant liquid to ultracentrifuge pot
5. Spin liquid at 28,000 x g 30 min (only if heavily soiled or high amounts of food debris)
6. Decant supernatant to ultracentrifuge tube
7. Spin supernatant at 240,000 x g
8. Discard supernatant and resuspend pellet

The efficiency of the procedure was tested by seeding samples in triplicate with 2 ml virus suspension in PBS, and resuspending the pellets in 2 ml cell culture medium M199 (Gibco BRL). Infectious viruses in seed suspensions and extract were enumerated by the method of Reed and Muench[5]. The efficiency of the protocol was 50 % for lettuce, 20 % for green onion, and 25 % for white cabbage. It was considered that the low efficiency of recovery from green onion may have been due to antiviral substances in the food.

2.1 Soft fruit

The following protocol has been developed for use with strawberries and raspberries. The sample sizes, 100 g and 90 g respectively, is determined as above.

1. Place sample in stomacher bag
2. Add 10 ml alkaline buffer
3. Stomach 1 min
4. Decant into centrifuge pot
5. Add 1 ml Catfloc TL[6]
6. Spin 28,000 x g 30 min
7. Decant supernatant to ultracentrifuge tube
8. Spin 240,000 x g 1 h
9. Discard supernatant and resuspend pellet

Often, a firm gel-like substance can be precipitated during ultracentrifugation, and the pellets can be difficult to solubilise. Consequently, the efficiency of infectious virus recovery varies between 5 and 65 % for strawberries and between 4 and 6 % for raspberries. These recoveries are nonetheless useful for survival studies and testing the effects of food processing upon viral contaminants. Effective methods of solubilisation are currently being tested.

3   CONCLUSION

The extraction protocols are rapid, taking only a few hours to perform, and are sufficiently efficient to allow their use in determination of virus survival on foods, or evaluating the effects of disinfectants and food processes on infectivity of viral food contaminants.
The methods require further refinement, to remove food debris and allow the viruses to be delivered to the detection system. We intend to employ a one-step extraction reverse transcription PCR technique[7], which will simplify the final procedures. Our ultimate aim is to be able to detect at least 10 virus particles per food sample.

ACKNOWLEDGEMENT

This work was supported by the Ministry of Agriculture, Fishery and Foods.

References

1. N. Cook. and S.H. Myint, *Rev. Med. Micro.*, 1995, **6**, 207 - 216.
2. S.H. Myint, 'Rapid Analysis Techniques in Food Microbiology', Blackie, Glasgow, 1995, pp. 170 -195.
3. Advisory Committee for the Microbiological Safety of Foods 'Foodborne Viral Infections', The Stationary Office, 1998.
4. Ministry of Agriculture, Fisheries and Food, 'Food Portion Sizes' 2nd Edition, HMSO, London, 1993.
5. L.J. Reed and H. Muench, *Am. J. Hyg.*, 1938, **27**, 493 - 497.
6. K.D. Kostenbader Jr. and D.O. Cliver, *Appl. Env. Micro.*, **41**, 318 – 320.
7. N. Cook, A.S. Kurdziel, J.P. Hays, S.H. Gordon, B.J. Donald, and S.H. Myint, S.H., *J. Rapid Meths Automat. Micro.*, **7,** 61 – 68..

CEN VALIDATION OF RAPID ALTERNATIVE MICROBIOLOGICAL
METHODS (MICROVAL)

R. Holbrook.

Convenor of Task Force 2,
CEN TC 275, Food Analysis - WG6 - Microbiology.
Retired from
Microbiology Section,
Unilever Research,
Colworth House,
Sharnbrook,
Bedfordshire, MK44 1PR, UK.

1 INTRODUCTION

MicroVal was set up in 1993 to create a single European mechanism for the
assessment of alternative (rapid) test kits for the microbiological analysis of foods
and food environment samples; but not water samples. It was recognised at that
time that manufacturers, or their agents, were frequently expected to support,
piecemeal, the assessment of their kits in many individual laboratories. This was
neither a cost or time effective process for the manufacturer. Invariably each
assessment was different and the quality of assessment varied according to the
expertise and capability of each laboratory. Standardisation of the test protocol via
the creation of a European CEN standard for the validation of qualitative and
quantitative alternative methods was intended. Also, the setting up of a MicroVal
Certification Scheme in Europe using Certification Bodies in several countries and
supervised centrally to MicroVal rules was envisaged. It was essential in order to
gain international acceptance and recognition, that the testing procedure gave the
reliable assessment of the alternative method's performance when compared against
ISO traditional methods, or there equivalent.

2 GENERAL

2.1 MicroVal Organisation

The MicroVal project gained Eureka status. At this stage the MicroVal group
was composed of individuals from 23 organisations from research, food
manufacturers, alternative kit producers, and standardisation institutes in Europe.
The Secretariat was the responsibility of Association Francaise de Normalisation

(AFNOR). Working groups had responsibility for the technical development and assessment of the test protocol, and elaboration of the Certification Scheme.

An alternative method of analysis may be the test procedure and reaction system, in whole or in part, proprietary or non-commercial that demonstrates or estimates the same parameter as is measured by the corresponding reference method. Through four years a protocol was developed and validated using four generically different methods. Two were qualitative methods - an enzyme linked immunoassay, and probe based method for the detection of two different foodborne pathogens, and two were quantitative methods, one for enumeration of Enterobacteriaceae and the other an automated impedance method for total viable count. Each was assessed against a corresponding ISO Standard for the microorganisms concerned. Qualitative methods assessed the relative accuracy, relative sensitivity, relative specificity, detection threshold, and relative detection level, together with an interlaboratory trial requiring 8 sets of results with no outliers from up to 15 laboratory trialists trained or experienced in the alternate and reference methods. Quantitative methods required the measurement of linearity, accuracy bias and relative accuracy of the alternative method compared with the reference method and then the determination of the relative limits of detection, quantification, critical level, relative specificity and selectivity against the reference method. An inter-laboratory trial to measure and compare the performance of reproducibility and repeatability was also part of the exercise. Eventually statistical methods for the analysis of all results was achieved.

2.2 CEN Standard

By 1997 the protocol was sufficiently well developed to request permission from CEN - European Committee for Standardisation, to produce a Standard based on the MicroVal work. This become the responsibility of a Task Force from Working Group 6 Microbiology, of Technical Committee 275 - Food Analysis. The Task Force comprised members of MicroVal and WG6 which were collectively from 9 European countries together with representatives from AOAC, AOAC RI, AFNOR, and DANVAL. Mutual recognition by other world wide bodies of the CEN Standard and the MicroVal certification process was desired. Currently a draft CEN Standard - prEN 275-061 Microbiology of food and animal feeding stuffs - Protocol for the validation of alternative methods, has been prepared. When approved the document will replace any national standards in European countries and also have ISO status through the Vienna Agreement.

The working draft comprises detailed protocols on all aspects of the validation process for both qualitative and quantitative methods. It includes the accreditation requirements of the organising laboratory, its role and the tasks it undertakes in the comparative evaluation of the alternative method against the reference method, the organisation and supervision of the inter-laboratory study. Rules for the selection of the reference method, acceptance of previous validations, selection of foods types appropriate to the alternative method, the use of naturally contaminated samples, preparation of spiked food samples, selection of test strains, rules on carrying out interlaboratory trials and assessment of data, together with other details, have been compiled into many annexes to the document. The methods of statistical analysis of all data is detailed together with some worked examples.

2.3 The MicroVal Certification Scheme

The data generated according to the CEN Standard 275-061 can be the basis for the certification of an alternative method by an independent certification organisation. Certification should ensure its acceptance by governmental inspection laboratories, other regulatory bodies and laboratories in the food trade, thus facilitating international commerce. The MicroVal organisation comprises a Secretariat at the Netherlands Normalisatie-instituut (NNI), a Group of Certification Bodies and a General Committee collectively responsible for expert and technical committees and a European network of approved expert laboratories, assessors and auditors. Independent certification organisations in France (AFNOR) and Germany are members working to the MicroVal rules for the certification of microbiological test kits for food analysis. Certification organisations in other European countries are being sought. The codes of practice are given in the MicroVal Rules and Certification Scheme, published by and obtainable from the MicroVal secretariat (1). It describes (amongst others) the specifications needed in an application file from the manufacturer, initial inspection, particular compliance of quality systems to ISO 90002 and audit to EN 29002 at the site of manufacture. It gives the requirements for labelling, certification and use of the licence. Certification fees are also stated that does not include the cost of the technical study. A flow chart and the time limits of each stage of the certification process for the manufacturer, MicroVal Certification Body, expert laboratory, inter-laboratory trial, method reviewers, statistical expert, auditors and production of the report are stated. The MicroVal Technical Committee and Certification Body make each certification decision. A MicroVal logo and the rules concerning its use by successful applicants are defined by the Certification Bodies. Requirements regarding documentation of any modifications to the alternative and reference methods and the regular inspection, (about every two years), of the quality of the certified method is needed to maintain the licence by the producer.

At this time the MicroVal Certification Scheme is still being set up for the validation of alternative methods of microbiological analysis for foods.

1) The MicroVal Secretariat, NNI, Postbus 5059, NL-2600 GB Delft, The Netherlands.

Rapid Methods in Food Chemistry

Professor K. Clive Thompson

The drive for rapid techniques in food analysis arises from the ever increasing pressure for positive release of products into the food chain, and a requirement for tighter control of the production process. This has spawned development of faster sample preparation and clean-up techniques, advances in rapid chromatography methods, rapid on-site single sample analysis and rapid screening methods. Robust low-cost at-site or on-site testing methods have considerable attraction to food companies.

Rapid screening methods using ELISA technology, have been designed to test food for the presence of a very wide range of both chemical and microbiological species. The chemical methods include species content of raw meats; uncooked processed foods; heated/canned meats; animal feed etc. ELISA test kits have also been used to determine protein additives/allergens (such as soya, peanut, wheat gluten, milk proteins, sesame etc); fungal toxins such as ochratoxin-A and aflatoxins; antibiotic residues such as sulphamethazine, tetracycline and streptomycin and more recently prion diseases such as BSE. Fit for purpose polyclonal and monoclonal antibody technology seems to be the key for successful ELISA kits.

Other exciting rapid method techniques for foods include various types of biosensors, DNA technology, NIR and chemometrics, surface plasmon resonance (SPR) and nuclear magnetic resonance (NMR) to analyse moisture and fat content of a food in less than 15 minutes with no sample preparation required.

However, no matter what rapid method (or test kit) is used, it is essential that it is fit for the intended purpose. The onus is on the user to ensure that it is suitable for their samples and it is strongly advised that some comparison is made with conventional laboratory analysis using a wide range of samples and some spiked samples before the new technique is adopted

The problems of taking a representative sample and transferring it without degradation to where the analysis is carried out should not be overlooked. Often significant attention is paid to the analysis stage whilst minimal effort is made to ensure that a truly representative sample is taken. Appropriate documented quality control checks need to be established before approving the use of any of these methods.

It is often erroneously assumed that minimal staff training is required with rapid methods and test kits. Without adequate documented training, poor quality data is likely to be produced. This key area is often overlooked and in the past has resulted in some unfair criticism of methods and test kits.

Proficiency testing is carried out by all responsible laboratories and is a key component in ensuring that the results being obtained are fit for purpose. However, there appears to be a reluctance to join proficiency schemes for testing carried out in non-laboratory locations. This attitude needs to be changed. Third party proficiency testing is a good way to demonstrate that the results are fit for purpose.

We are living in a time of rapid change and it will be interesting to see how much routine analysis moves out of the laboratory to the production line or site in the next ten years. It is hoped that this book will give the reader some ideas of the future direction of rapid food analysis methods.

BIOSENSORS FOR PESTICIDE RESIDUES IN THE ENVIRONMENT AND IN FOOD

Petra M. Krämer

Institute of Ecological Chemistry
GSF - National Research Center for Environment and Health
Ingolstädter Landstr. 1
85764 Neuherberg
Germany

1 INTRODUCTION

Classical analytical methods, e.g. LC (liquid chromatography) and GC (gas chromatography) with their various detection systems are widely used and accepted for pesticide residue analysis, including major pesticide metabolites and degradation products. However, these analytical techniques are time consuming and expensive, they need mostly extensive sample preparation and preconcentration, sometimes even derivatization. Only well equipped laboratories have sufficient resources to carry them out on a routine basis, and they are difficult to adapt for field use or on site testing. As a result, there is a need for alternative and/or complementary methods, such as biochemical analysis and biosensors.

The process of obtaining inexpensive, fast, sensitive, reliable and standardized biosensors, biochemical techniques and automated devices for either broad-spectrum or single analyte selective analysis is still ongoing. However, during the past five years there has been a notable progression in this direction.[1,2] Examples given in this paper are undoubtedly only a small fraction of the biosensor research in environmental and food analysis and focus mainly on pesticide analysis. However, these examples, e.g. acetylcholinesterase inhibition test,[3] immunosensors and automated flow injection immunoanalysis,[e.g. 4,5] whole cell biosensors,[6] receptor based assays,[7] as well as different transducer techniques[8] represent major research fields. They are meant to serve as an useful starting point to get information about trends and they should encourage more future research and development in these interdisciplinary fields.

2 DIFFERENT BIOCOMPONENTS IN BIOSENSORS

Biosensors consist of a biological component in close relationship to a transducer. The selectivity of a biosensor is determined by the biomolecule or biological element, which in most cases is much more selective and in addition more sensitive than any other material (e.g. molecular imprinted polymers, ion selective electrodes). Generally, a differentiation between biocatalytic and bioaffinity sensors is made. Major problems with these biological components though are still their lack of stability, the required selectivity for the chosen purpose, and very often lack of commercial availability and standardization.

Nevertheless, a lot of different biological recognition elements have been used in different biosensor set-ups. The main elements were selected here and some examples were chosen to demonstrate their principles.

2.1 Biosensors Based on Enzymes

Enzymes used for pesticide determination are not as widely used as enzymes in medical and pharmaceutical applications or in fermentation control. In medical diagnostics, glucose oxidase is certainly the most successfully used enzyme in biosensor development and commercialization. In environmental analysis, and here particularly in pesticide residue analysis, only very few enzymes are used, e.g. urease, tyrosinase and especially butyryl- and acetylcholinesterases.

2.1.1 Acetylcholinesterase (AChE) Inhibition. AChE has been utilized either alone or in combination with cholineoxidase and the reaction was transduced with nearly all available principles. A selected list of some major developments is presented in Table 1.

Table 1 *Selected Biosensors Based Upon AChE Inhibition (and one Other Principle) for Organophosphorus and Carbamate Pesticides - Determination with Different Transducer Principles*

Enzyme	Concentration Determined (Detection Limit)*	Matrix	Transducer	Reference
AChE	Dichlorvos: $7\text{-}10^{-11}$ mol l^{-1} Paraoxon: 4×10^{-11} mol l^{-1}	Buffer, River water	Screen printed amperometric electrode	9
AChE	Paraoxon: 5×10^{-8} M, Carbaryl: 1×10^{-7} M (after 5 min pre-incubation)	Buffer	Quartz crystal microbalance, indigo pigment precipitation	10
AChE	Carbofuran: 1.5×10^{-8} M Paraoxon: 1.1×10^{-7} M	Buffer	Fiber optic, thymol blue indicator	11
AChE	Carbofuran: sub ppb level	Buffer	Screen printed pH electrode	12
E. coli with-organophos-phorus hydrolase (OPH)	Paraoxon: 3 µM Parathion: 3 µM Coumaphos: 5 µM	Buffer (Waste-water)	Fiber-optic bundle	13

* Units are presented as given in the reference

2.2 Biosensors Based on Antibodies

Antibodies have been developed during the last 20 years for many different pesticides.[14] Most of these immunoreagents have been developed for conventional ELISA (enzyme linked immunosorbent assay), but have also been used in immunosensors.

2.2.1. Immunosensors. Immunosensors can be divided in direct and indirect measuring devices. Direct immunosensors need no additional reagent for the measurement of the antigen-antibody reaction (ag-ab-reaction). Examples for transducers in these set-ups are devices based on piezoelectric effect, evanescent wave, and surface plasmon resonance. Generally, these formats are easier and faster in their performance, but the sensitivity reached is in many cases not high enough, and unspecific reactions are harder to control. Until today, the level of sensitivity is still the highest, when a competitive (indirect) format is used, which means a label or tracer is needed; one important example is drinking water analysis, where a determination at the 0.1 μg l^{-1} level or lower is needed. For other purposes, for example when fruit extracts have to be analyzed, a sample preparation and also higher concentrations make direct measuring principles suitable. Table 2 presents some selected examples of immunosensors with different transducers, which have been realized in the past few years.

Table 2 *Selected Immunosensors With Different Transducers*

Analyte Detected	Concentration Determined (Detection Limit)*	Matrix	Transducer	Reference
- African swine fever virus - Ab 18BG3	1 μg ml^{-1} 0.2 μg ml^{-1}	Diluted pig serum	Quartz crystal microbalance (bulk acoustic wave)	15
Atrazine	3 μg l^{-1}	Dist. water, water samples	Conducting polymer-coated platinum electrode	16
2,4-D	1 μg l^{-1}	River water, lake water	Optical sensor chip, evanescent wave	17
2,4-D	0.01 μg l^{-1}	Tap water	Screen-printed amperometric electrode	18
2,4-D	5 μg l^{-1}	Buffer, surface water	Potentiometric, LAPS (light addressable potentiometric sensor)	19

* Units are presented as given in the reference

2.2.2. Automated Flow Injection Immunoanalysis. A further step in immunoanalysis is automation. In contrast to biocatalytic enzyme reactions, the ag-ab-reaction is not an on-off-reaction and usually washing steps are needed to separate bound from unbound material. Based on this, it is very useful to automate the association and dissociation of the ag-ab-reaction and the assay steps. This is mostly done by flow injection analysis. Many developments realized this automation during the last few years. These approaches are either focusing on the regeneration of the ag-ab-reaction or on the regeneration of a column support with protein A or G. The later is more promising, because it is a more universal binding reaction and therefore optimizations are less complex. A list of some examples is given in Table 3. Here, the main focus is on liquid samples, which make it possible to use these systems directly without sample pretreatment.

Table 3   *Selected Automated Immunosensors Combined with Flow Injection Systems*

Analyte Detected	Concentration Determined (Detection Limit)*	Matrix	Transducer	Reference
Atrazine	2.1 μg l^1	Drinking water, orange juice	Fluorescence detector	20
Atrazine	---	---	Ag/AgCl electrode with HRP	21
Atrazine	9 μg l^1	Buffer	Amperometric detector	22
Simazine	0.2 μg l^1	Ground- and surface water	WSPR (wave-guide surface plasmon resonance)	23
Diuron	0.02 μg l^1	Drinking water supplies	Fluorescence detector	24
Irgarol 1051	0.01 μg l^1	Beach seawater, river / lake water	Fluorescence detector	25
Carbaryl	0.029 μg l^1	Drinking water, apple juice	Fluorescence detector	26
13 Analytes planned (e.g. atrazine, PCP, glyphosate, isoproturon, etc.)	1 μg l^1 or below	River water	Optical waveguide, TIRF (total internal reflection fluorescence)	27

* Units are presented as given in the reference

2.3 Biosensors Based on Receptors

Receptor binding assays are an approach of biological effect-related analysis. This is a distinction to immunoanalysis, where the binding of the antibody to the antigen has no relation to an effect. Examples for receptor based assays used in environmental analysis are the PS II reaction center for the detection of PSII herbicides, such as atrazine, diuron, ioxynil, etc., or the usage of human estrogen receptors for the detection of estrogenic substances[7] in the environment.

2.4 Biosensors Based on Organelles or on Whole Cells

For the determination of pesticides, biosensors within this group are mainly focusing on the determination of PS II herbicides (s-triazines, phenylurea herbicides). They are either based on organelles such as chloroplasts (thylakoids)[28] or on whole cells like cyanobacteria (with mediator)[29] or unicellular algae. Whole cells have the advantage of stability, but the sensitivity of the analysis is too low for some applications. In control situations though, where only a screening is needed, e.g. in waste water treatment plants, some biosensors based on whole cells are already routinely used.

Generally, the problem with organelles is a) that they are not commercially available and b) that the standardization is missing.

3 TRANSDUCERS

Transducers used in biosensors for pesticide residue analysis are generally not different from other application fields. The detection of the interaction of the biological molecules cause a change in physico-chemical parameters, e.g. ions, electrons, gases, heat, mass or light. Optical systems are very widespread, because enzyme reactions are easy to determine with these transducers. An overview on optical probes and transducers is given for example by Brecht and Gauglitz.[30] The evanescent field and the surface plasmon resonance are two optical phenomena which are certainly very useful for many applications. Developments with major commercial success are for example based on surface plasmon resonance (e.g. BIAcore[TM], Biospecific Interaction Analysis, Pharmacia). Although the main usage of these systems is in pharmacy or medicine, there are some groups, who also used this system for pesticide analysis.[e.g. 31] In addition, the application of screen-printed electrodes as electro-chemical transducers is often realized, because these electrodes can be easily produced very cheap in high quantities.

A very important and highly promising approach in transducer developments is the miniaturization (biochip, microchip) with the aim of multi-analyte analysis for monitoring continuously key analytes in the environment. This will enable in the future the effective control of different environments such as water ways, ground water or effluents of pesticide production plants.

4 SUMMARY AND CONCLUSION

Although many developments are currently in progress, there is still a lack of acceptance in the analytical field. Usually, analytical chemists are used to conventional analysis and use standard operating procedures, which are established in their laboratories. In addition, the level of reliability will determine the method that is allowed to be used. Here, biosensors are still lacking quality assurance and standardization and are not used as a routine method.

A common problem in food and environmental control and analysis is, that sample preparation is often needed. When this is the time limiting step in analysis, the time savings in biosensor analysis is not of significance any longer. In this respect, only the savings of solvents, automation of analysis, and money savings in the instrumental costs favor biosensors or biochemical analysis.

Big advantages are coming into place though, when no sample preparation is needed, e.g. in water analysis and control, and here an automated biosensor system, which is stable for 1-2 weeks is certainly a very good and useful complementary to conventional analysis of today.

An additional interest is also observed in the integration of biochemical methods or biosensors with conventional analysis, e.g. integration of LC with immunoassay, such as immunoaffinity columns prior to LC or LC-MS-MS, and biosensor detection following LC.

REFERENCES

1   P. M. Krämer, *J. AOAC Int.*, 1996, **79 (6)**, 1245.
2   P. M. Krämer, *Inside Laboratory Management*, 1998, January, 23.
3   P. Skladal, *Food Technol. Biotechnol.*, 1996, **34 (1)**, 43.
4   E. P. Meulenberg, *Food Technol. Biotechnol.*, 1998, **35 (3)**, 153.
5   P. M. Krämer, B. A. Baumann and P. G. Stoks, *Anal. Chim. Acta*, 1997, **347**, 187.
6   M. Naessens and C. Tran-Minh, *Anal. Chim. Acta*, 1998, **364**, 153.
7   M. Seifert, S. Haindl and B. Hock, 'Advances in Experimental Medicine and Biology', 444, del Mazo (Ed.), Plenum Press, New York, 1998, pp. 113.
8   R. S. Sethi, *Biosens. Bioelectron.*, 1994, **9**, 243.
9   J. J. Rippeth; T. D. Gibson, J. P. Hart, I. C. Hartley, G. Nelson, *Analyst*, 1997, **122 (11)**, 1425.
10   J. M. Abad, F. Pariente, L. Hernandez, H. D. Abruna, E. Lorenzo, *Anal. Chem.*, 1998, **70 (14)**, 2848.
11   R. T. Andres, R. Narayanaswamy, *Talanta*, 1997, **44 (8)**, 1335.
12   R. Koncki, M. Mascini, NATO ASI Ser., Ser. 2, 1997, 38 ('Biosensors for Direct Monitoring of Environmental Pollutants in the Field') Kluwer Academic Publishers, pp. 139.
13   A. Mulchandani, I. Kaneva, W. Chen, *Anal. Chem*, 1998, **70 (23)**, 5042-5046.
14   E. P. Meulenberg, W. H. Mulder, P. G. Stoks, *Environ. Sci. Technol.*, 1995, **29 (3)**, 553.
15   E. Uttenthaler, C. Kößlinger, S. Drost, *Biosens. & Bioelectron.*, 1998, **13**, 1279.

16   T. L. Fare, M. D. Cabelli, S. M. Dallas, D. P. Herzog, C. S. Hottenstein, T. S. Lawruk, F. M. Rubio J.C. Silvia, S. S. Wiedman, 'Field Screening Methods Hazard. Wastes Toxic Chem.', Proc. Int. Symp., 1995, Vol. 1, 109-120, Air & Waste Management Association, Pittsburgh, PA, USA.

17   M. Meusel, D. Trau, A. Katerkamp, F. Meier, R. Polzius, K. Cammann, *Sens. Actuators B,* 1998, **51**, 249.

18   T. Kalab, P. Skladal, *Electroanalysis*, 1997, **9** (4), 293.

19   L. Piras, M. Adami, S. Fenu, M. Dovis, C. Nicolini, *Anal. Chim. Acta*, 1996, **335**, 127.

20   E. Turiel, P. Fernandez, C. Perez-Conde, A.M. Gutierrez, C. Camara, *Talanta*, 1998, **47 (5)**, 1255.

21   R. W. Keay, C.J. McNeil, *Biosens. Bioelectron.*, 1998, **13 (9)**, 963.

22   F. Vianello, L. Signor, A. Pizzariello, M. L. Di Paolo, M. Scarpa, B. Hock, T. Giersch, A. Rigo, *Biosens. Bioelectron.*, 1998, **13 (1)**, 45.

23   C. Mouvet, R. D. Harris, C. Maciag, B. J. Luff, J. S. Wilkinson, J. Piehler, A. Brecht, G. Gauglitz, R. Abuknesha, G. Ismail, *Anal. Chim. Acta*, 1997, **338**, 109.

24   P. M. Krämer, *Food technol. biotechnol.*, 1998, **36 (2)**, 111.

25   M. A. González-Martínez, J. Penalva, R. Puchades, A. Maquieira, B. Ballesteros, M. P. Marco, D. Barcelo, *Environ. Sci. Technol.*, 1998, **32 (21)**, 3442.

26   M. A. González-Martínez, S. Morais, R. Puchades, A. Maquieira, A. Abad, A. Montoya, *Anal. Chem.*, 1997, **69**, 2812.

27   C. Barzen, A. Brecht, I. Stemmler, G. Gauglitz, R. Abuknesha, Poster at the EU-Workshop 'Biosensors for environmental monitoring', Freising, Germany, 28-30 May, 1997

28   D. Merz, M. Geyer, D. A. Moss, H.-J. Ache, *Fresenius J. Anal. Chem.*, 1996, **354** (3), 299.

29   D. M. Rawson, A. J. Willmer, A. P. F. Turner, *Biosensors*, 1989, **4**, 299.

30   A. Brecht, G. Gauglitz, *Biosens. Bioelectron.*, 1995, **10**, 923.

31   M. Minunni. M. Mascini, *Anal. Lett.*, 1993, **26 (7)**, 1441.

DEVELOPMENT OF DISPOSABLE IMMUNOSENSORS FOR THE RAPID ASSAY OF SEAFOOD TOXINS

L. Micheli, D. Moscone, S. Marini*, S. Di Stefano and G. Palleschi

Dipartimento di Scienze e Tecnologie Chimiche, Universita' di Roma "Tor Vergata", Italy
*Dipartimento di Medicina Sperimentale e Scienze Biochimiche, Università di Roma "Tor Vergata" Italy
tel. 39-6-72594394 Fax 39-6-72594328 E-mail: Moscone@uniroma2.it

1 INTRODUCTION

Saxitoxin and domoic acid are potent marine toxins involved respectively in paralytic shellfish poisoning (PSP) and in anamnestic shellfish poisoning (ASP). Human ingestion of seafood contaminated with these toxins resulted, in fact, in severe intoxication and occasionally death. Because of the potential health hazard, there is the need for quick, sensitive and specific toxin detectors suitable also for field use. Conventional mouse bioassays, used for the detection of most of these toxins, cause problems both ethical, because of the use of living animals, and technical, because of its poor specificity and sensitivity. HPLC methods are more sensitive than the mouse bioassay, but require expensive instrumentation, chemical clean-up steps and skilled operators. Immunosensors offer the possibility of rapid, inexpensive, easy to perform assays, particularly as rapid screening tests.

The production of disposable immunosensors for toxin assays is the subject of this work, according with the following steps: production of polyclonal and/or monoclonal antibodies against selected seafood toxins; development of the enzyme-toxins conjugates with peroxidase enzyme; production and characterisation of disposable electrodes for enzyme activity measurements and toxin assays. Polyclonal antibodies against saxitoxin and domoic acid have been obtained by immunisation of mice and rabbits. Conditions for the immunoassays have been set-up by spectrophotometric Enzyme-Linked ImmunoSorbent Assay (ELISA) tests.

The 3,3',5,5'-tetramethylbenzidine (TMB), used as enzyme-substrate, allows a high sensitive procedure of analysis (spectrophotometric and electrochemical).

2 MATERIALS AND METHODS

2.1 Production of polyclonal antibodies

Four-week old female Balb/c mice were immunised using saxitoxin or domoic acid conjugated to *Recombinant Hepatitis B core Antigen* by intraperitoneal injection with complete Freund's adjuvant. Subsequent boosts were repeated every 3-4 weeks by using incomplete Freund's adjuvant and phosphate buffer solution (PBS). Sera were harvested 7 days after last boost, stocked and stored at -20°C until use.

2.2 ELISA procedures for saxitoxin

Sera were screened for the presence of specific antibodies by non competitive spectrophotometric ELISA tests. Several parameters have been optimised in order to obtain the best signal/noise ratio and the highest sensitivity.

The selected conditions were as follows: 50 µl of a saxitoxin solution 9 µM in EtOH were adsorbed on 96-wells plates (Maxisorp, Nunc), for 2 hours at 37 °C. To reduce unspecific binding, 1% skimmed milk in PBS (50 µl/well) at 37 °C for one hour was then added to each well followed, after three washings with PBS/0.01% Tween 20, by the addition of antisera dilutions (50 µl per well). After 2h at 4°C, unreacted primary antibody was washed out by three washings with PBS/0.01% Tween 20. Binding of antibodies was then detect by colour development upon addition of 50 µl/well rabbit anti mouse IgG conjugated with Horseradish Peroxidase (HRP) enzyme as secondary antibody. After a suitable time, plates were washed with PBS/0.01% Tween 20 and TMB substrate was added to detect the residual peroxidase activity depending on the amount of trapped antibodies. The colorimetric reaction was stopped by adding 50 µl/well 2 M HCl. Absorbances at 492 nm were determined for each well with a microplate reader (model 550, Bio-Rad).

Using the sera harvested after 5-7 cycles of immunisation, a competitive assay has been set-up. Different concentrations of saxitoxin were mixed with a 1:1000 dilution of the antiserum and added to saxitoxin-coated plates. After 1 h at 4°C plates were washed three times with PBS/0.01% Tween 20; secondary antibody was added and left to react for 30 min at 4°C. Residual peroxidase activity was then detected as previously described.

2.3 ELISA procedures for domoic acid

The presence of the polyclonal antibodies against domoic acid (DA), was showed by spectrophotometric ELISA tests. Results demonstrated that the better conditions were obtained using, as coating solution, domoic acid conjugated by EDC to *Hemocyanin* from *Keyhole Limpets* (KLH);[1] this conjugate was used at a concentration of 1 µg/mL in carbonate buffer 0.1 M pH 9.6 and left overnight at 4°C on Falcon 3912 plates. 1% skimmed milk plus 10 µg/mL IgG (rabbit) in PBS at 37°C for 1 h was used as blocking solution. The competitive curve was obtained mixing DA, at different concentrations, with 1:1000 dilution of the antiserum and incubating the mixture for 12 min at 4°C onto DA-KLH coated plates. After washings and secondary antibody addition, residual peroxidase activity was then detected as previously described.

3 RESULTS AND DISCUSSION

Figures 1 and 2 show the saxitoxin and domoic acid binding curves. From these results the appropriate antisera dilution to be used in subsequent competitive tests was calculated; in our case a dilution 1:1000 has been selected for both antisera. Figures 3 and 4 report the competitive curve obtained; the results obtained indicate that the antibodies against saxitoxin react with this toxin approximately in the range 5×10^{-7}-5×10^{-5} mol/L, while for domoic acid the useful range is between 10^{-9}-10^{-6} mol/L. Moreover, from preliminary results, a mean dissociation constants of about 10^{-7}-10^{-6} mol/L and 10^{-8}-10^{-7} mol/L can be calculate for saxitoxin and domoic acid antibodies respectively. These values are quite promising for the development of the seafood sensors.

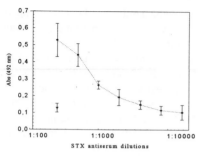

Figure 1 *STX binding curve*

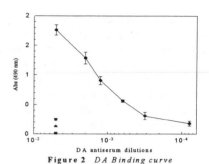

Figure 2 *DA Binding curve*

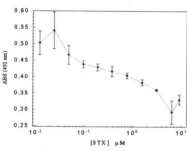

Figure 3 *STX competition curve*

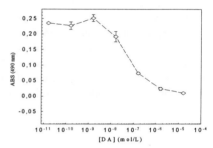

Figure 4 *Domoic Acid competitive curve*

4   ELECTROCHEMICAL   SENSORS   FOR   ENZYME   ACTIVITY   DETERMINATION

In a previous work,[2] the use of 3,3',5,5'-tetramethylbenzidine (TMB) as electrochemical substrate of horseradish peroxidase enzyme (HRP) has been investigated. HRP activity has been detected in flow injection analysis (FIA) in citrate-phosphate buffer, pH 5, using a glassy carbon as working electrode, poised at a potential of +100 mV *versus* an Ag/AgCl as reference and with stainless steel as auxiliary electrode.

Optimised substrate concentrations were 2×10^{-4} mol/L for TMB and 10^{-3} mol/L for H_2O_2, and the detection limit after 15 minutes of incubation was 8.5×10^{-14} mol/L for HRP with the amperometric method.

4.1 Characterisation of the Screen Printed Electrodes (SPE)

The screen-printed electrodes were supplied by the Biosensor Laboratory of the University of Florence (Italy). The printing procedure is reported elsewhere.[3] Briefly, the electrochemical strips consisted of a carbon working electrode and two silver pseudoreference and counter electrodes, all screen-printed on a flexible polyester film.

Table 1 *Response of 5 different electrodes (mean values ± SD) to addition of ferricyanide 2 mM. (pretreatment=90 sec at 1 V)*

	triple electrodes
Before polarisation	-0.80 ±0.11 µA RSD= 14%
After polarisation	-0.99 ±0.08 µA RSD= 8%

For their characterisation, cyclic voltammograms of different electrode configurations were carried out, in presence or in absence of external reference and auxiliary electrodes and in different buffers. Results showed the presence of a signal noise in the reduction region especially in presence of in the reaction medium of KCl. This noise, also reported in literature for similar electrodes,[4] did not interfere with the measurement of the TMB substrate.

The reproducibility of the screen printed electrodes was evaluated using $K_3Fe(CN)_6$ as model substrate because of the no commercially availability of the oxidised form of the TMB. Also polarisation pretreatments, reported in literature to activate the screen printed electrodes,[5] were performed, an the results are reported in the table 1. Results are in agreement with the literature, showing an increase of the sensitivity and of the reproducibility of the electrochemical measurements.

4.2 HPR Activity Measurements with SPE

These measurements were performed using TMB, H_2O_2 and the enzyme free in solution. Results showed an useful range of enzyme activity measurements of $2.5x10^{-5} \div 10^{-3}$ U/ml, after an incubation time of 5 minutes.

Similar experiments were repeated with the enzyme previously absorbed on the disposable electrodes surface, in order to mimic the final conditions, where the enzyme will be in close contact to the electrode surface, linked to the antigen or to the antibody. In this case the sensitivity was higher than the previous results, with a linear range of activity between $5x10^{-6} \div 10^{-3}$ U/ml (r^2= 0.987, n=3).

Future work will be focused on the use of different electrochemical techniques in order to improve the sensitivity of the enzyme activity measurements.

This work was supported by the financial contribution of the EU-FAIR CT 95-1092 research project.

References

1. H. Newsome, J. Truelove, L. Hierlihyand, P. Collins *Bull. Environ. Contam. Toxicol.*, 1991, **47**, 329-334.
2. G. Volpe, D. Compagnone, R. Draisci and G. Palleschi *Analyst*, 1998, **123**, 1303-1307.
3. M. Del Carlo, I. Lionti, M. Taccini, A. Cagnini and M. Mascini *Anal. Chim. Acta*, 1997, **342,** 189-197.
4. Kroger and A.P.F. Turner, 1997, *Anal. Chim. Acta, 347*, 9-18.
5. Wang, M. Pedrero, H. Sakslund, O. Hammerich *Analyst*, 1996, **121,** 345-350.

SELECTIVE ELECTROCHEMICAL BIOSENSORS FOR APPLICATION IN FOOD QUALITY CONTROL

G. Palleschi, D. Compagnone and D. Moscone.

Dipartimento di Scienze e Tecnologie Chimiche, Università di Roma "Tor Vergata", Italy
tel. 39-0672594843 fax 39-0672594328 e-mail: Palleschi@stc.uniroma2.it

1 INTRODUCTION

The demand for sensitivity, specificity, speed and accuracy of analytical measurements has stimulated considerably the interest in developing electrochemical probe procedures as diagnostics in food technology. Measurement of quality marker analytes in food are routinely performed with chromatographic and spectroscopic techniques. Electrochemical biosensors offer a valuable alternative with advantages in rapidity, handling of reagents, sample throughput rate and cost effectiveness for the assay of a great number of these food components. Moreover, they can be easily configured for flow-through or flow-injection analysis (FIA) allowing a real-time monitoring for process control. In this work we report on the measurement of glucose and fructose in honey samples, lysine in food products, biogenic amines in stored fish samples, malic acid and glycerol in wines and lactulose in milk using amperometric biosensors which resulted suitable for the assessment of the quality of food products.

1.1 A tri-enzyme electrode probe for the sequential determination of fructose and glucose in honey samples

A new probe for the determination of fructose and glucose has been assembled using a platinum electrode and three enzymes co-immobilised on its surface: fructose dehydrogenase (FDH), glucose dehydrogenase (GDH) and diaphorase[1]. The mechanism of this probe is very simple: when the probe is immersed in a solution containing fructose and glucose and an electrochemical mediator, the current signal is due to the fructose. Then, by adding $NAD(P)^+$ a current response due to the glucose present in solution is obtained. The overall reaction scheme is the following:

$$\text{fructose} + \text{mediator}_{ox} \xrightarrow{\textbf{FDH}} \text{chetofructose} + \text{mediator}_{red}$$

$$\text{glucose} + NAD^+ \xrightarrow{\textbf{GDH}} \text{gluconolactone} + NADH$$

$$NADH^+ + \text{mediator}_{ox} \xrightarrow{\textbf{diaphorase}} NAD^+ + \text{mediator}_{red}$$

Table 1. *Analysis of fructose and glucose in honey samples using the biosensor (B) and a spectrophotometric procedure (S).*

	%fru (w/w)			%glu (w/w)		
n.	B	S	E (%)	B	S	E (%)
1	40.0	39.5	1.3	33.0	35.0	5.7
2	33.0	35.0	5.7	26.0	24.0	8.3
3	33.0	35.0	5.7	31.0	31.0	0.0
4	37.0	37.5	1.3	26.0	27.0	3.7
5	32.0	34.0	5.9	27.0	24.0	12.5
6	38.0	39.0	2.6	27.5	27.0	1.8
7	33.0	35.0	5.7	28.0	28.0	0.0
8	35.0	37.0	5.4	22.0	23.0	4.3
9	39.0	40.0	2.5	31.0	29.0	6.9
10	40.0	41.0	2.4	35.0	35.0	0.0

Ferricyanide (3 mmol/L) was used as mediator. Calibration curves for fructose and glucose have been constructed. The linear range for both fructose and glucose was from 5×10^{-6} to 2×10^{-4} mol/L and the detection limit was 1×10^{-7} mol/L.

The probe has been used to determine the concentration of fructose and glucose in ten different samples of honey. Commercial honey contains fructose and glucose at percentages of ~ 40% and ~ 30%, respectively.

The analysis of these two compounds in honey is at present carried out using HPLC or spectrophotometric procedures. The amperometric method was more reproducible than the other: the average RSD of the ten samples tested was 3.1 % for the amperometric measurements and 4.4 % for the spectrophotometric (Table 1). Correlation was good.

1.2 Electrochemical bioprobes for lysine coupled with conventional and microwave protein hydrolysis

The determination of proteic lysine is an important parameter for the quality of foods and feeds. In fact, being lysine an essential aminoacid (not sintethized by mammals) the total lysine content is used as index of the nutritional quality of food.

Amperometric enzyme electrode probes have been constructed for the specific determination of L-lysine and used in batch and flow analysis[2]. The enzyme lysine oxidase was immobilised on a preactivated polymer support which was placed on a platinum electrode. Additional blocking membranes conferred high stability, reproducibility and avoided electrochemical and enzyme interferences. Parameters including pH, temperature, storage and operational times were optimised. The reaction is as follows:

$$\text{L-lysine} + O_2 + H_2O \rightarrow \alpha\text{-keto-}\varepsilon\text{-aminocaproate} + NH_3 + H_2O_2$$

Table 2. *Comparative study between lysine values (g/100g protein) obtained by the combinations of traditional and biosensor procedures.*

S	A	B	C	D	A/B	A/D	B/C	C/D	B/D
					— E (%)				
BSA	11.23	12.35	11.75	11.65	9.9	4.6	4.9	1.6	6.4
Farro whole meal		3.14	3.13	3.06			0.3	2.2	2.5
Farro sieved meal		2.55	2.80	2.82			9.8	0.7	10.6

S = sample, A = theoretical value, B = traditional hydrolysis + IEC, C = microwave hydrolysis + IEC, D = microwave hydrolysis + biosensor

The production of H_2O_2 is detected by the platinum electrode held at 650 mV applied potential *vs.* a silver /silver chloride cathode. The output current is correlated to the concentration of lysine in the sample.

Analysis in feeds is usually carried out by acid hydrolysis to liberate lysine; then the solution analysed with the bioprobe and HPLC procedures. The overall analysis takes more than 24 h.

More recently a fast procedure for lysine analysis in food was developed by coupling in sequence a microwave protein hydrolysis technique with the lysine enzyme electrode[3].

Protein hydrolysis was carried out in 6 N HCl using sealed vessels located in a microwave digestion system. Parameters such as irradiation power, pressure, time and temperature were varied to select the best conditions for the hydrolysis.

Bovine serum albumin and two farro (*T. dicoccum*) meals (whole and sieved) were processed for lysine content. The analysis was carried out with this new procedure (microwave hydrolysis + biosensor) and by traditional hydrolysis coupled with ion-exchange chromatography (IEC). Results obtained with the two procedures are reported in Table 2.

The microwave-biosensor procedure allows measurement of lysine in 30-60 min versus the 24-48 h of the traditional hydrolysis-HPLC procedure.

1.3 Salted anchovies storage conditions measured using a biogenic amine biosensor

The assembling and optimisation of an electrochemical biosensor for the determination of biogenic amines (putrescine, cadaverine, histamine, tyramine, spermidine, spermine, tryptamine) commonly present in food products, and its application to salted anchovy samples has been realised[4].

Variations of the amine content in anchovies during storage time were measured both with the biosensor and ion chromatography with integrated pulsed amperometric detection (IC-IPAD). The probe is based on a platinum electrode which senses the hydrogen peroxide produced by the reaction catalysed by the enzyme diamine oxidase (DAO), purified from commercial seeds of cicer and immobilised on the electrode surface. The generic reaction is as follows:

$$R\text{-}CH_2\text{-}NH_2 + O_2 \rightarrow R\text{-}CHO + H_2O_2 + NH_3$$

The content of amines in samples was measured as follows: the assembled biosensor was allowed to equilibrate in 3 mL of phosphate buffer (pH 8) at room temperature and under magnetic stirring until a steady current baseline was reached. This took about 10 minutes. Aliquot of samples (200 µl) or amine standard solutions were then injected and a current change due to the H_2O_2 production was recorded and related to the amine concentration.

Parameters as enzyme immobilisation and pH have been studied in order to obtain similar sensitivity for all the amines tested. Immobilisation of the enzyme on a nylon-net membrane, using glutaraldeyde as cross-linking agent, and phosphate buffer at pH 8.0 were used. The detection limit was 5×10^{-7} mol/L. The linear range common to the amines tested was obtained from 1×10^{-6} mol/L to 5×10^{-5} mol/L. The effect of potential interfering compounds was also evaluated. Underivatised biogenic amines such as putrescine, cadaverine, histamine, tyramine and spermidine were also detected with the IC-IPAD method.

Changes in the concentration of biogenic amine content in salted anchovy samples measured with the biosensor and IC-IPAD method exhibited the same trend and demonstrated that the biosensor is an useful tool to monitor the variation of the total amine content in fish during storage.

1.4 Analysis of Malic Acid in wine and must samples

A new amperometric malate enzyme electrode probe has been also constructed using a hydrogen peroxide based sensor coupled with malic enzyme and pyruvate oxidase[5].

The reactions are as follows:

$$L\text{-malate} + NADP^+ \leftrightarrow pyruvate + NADPH + CO_2 + H^+$$

$$Pyruvate + HPO_4^{2-} + O_2 \leftrightarrow acetylphosphate + H_2O_2 + CO_2$$

The malic enzyme catalyses the oxidation of malic acid, which in the presence of $NADP^+$ yields pyruvate as product. The oxidation of pyruvate is catalysed by pyruvate oxidase, which yields H_2O_2 as product in the presence of O_2 and phosphate as cosubstrates and thiamine pyrophosphate and Mg^{2+} as cofactors. The H_2O_2 is then detected by the electrochemical transducer , and the output current changes are correlated to the concentration of malic acid in solution. Analytical parameters such as pH, temperature, buffer, substrate and cofactor concentrations, and response time have been optimised. Coimmobilisation of the oxidase and dehydrogenase enzymes has been performed both randomly and asymmetrically on different supports. Calibration curves for malate have been constructed with all the analytical parameters optimised. The detection limit for this newly designed probe was 5×10^{-7} mol/L, with a broad linear range between 10^{-6} and 5×10^{-4} mol/L.

Recovery tests of malate in a wine matrix showed a percentage of recovery ranging from 96 to 101 % successfully. Malic acid has been determined in grape musts during grape maturation. Results, reported in Table 3, correlated well with those from a spectrophotometric procedure.

The results obtained indicate that both lactic and malic acids can be accurately measured with the biosensors. The rapidity of the analysis (about 2 min) allows on-line monitoring of malolactic fermentation during wine production.

Table 3. *Spectrophotometric (S) and amperometric (A) analysis of malic acid in must samples during maturation of grapes.*

		Amount	found	(g/L)
Sample	Day/month	S	A	Rel error (%)
Trebbiano	07/09	8.75	8.65	1
Trebbiano	13/09	6.05	5.85	3
Trebbiano	22/09	4.55	4.53	0
Trebbiano	28/09	2.57	2.83	10
Tintilia	08/09	4.54	4.29	6
Tintilia	14/09	4.12	4.28	4
Tintilia	22/09	3.52	3.57	1
Tintilia	28/09	3.05	2.65	15
Tintilia	05/10	2.38	2.15	10

1.5 Glycerol

Glycerol is the most important secondary product of alcoholic fermentation and contributes to the smoothness and viscosity of a wine with a favourable effect on the taste[6]. The amount of glycerol formed in wines ranges from 1 to 10 g/L

We developed an amperometric FIA biosensing system for the determination of glycerol[7]. The reaction scheme is the following:

$$\text{Glycerol} + \text{ATP}(\text{Mg}^{2+}) \longrightarrow \text{Glycerol-3-phosphate} + \text{ADP}$$
$$\text{Glycerol-3-phosphate} + O_2 + H_2O \longrightarrow \text{Glycerone-3-phosphate} + H_2O_2$$

The first reaction is catalysed by glycerokinase (GK), the second by glycerol-3-phosphate oxidase (GPO). A platinum based H_2O_2 probe polarised at +650 mV vs. Ag/AgCl was used as electrochemical transducer.

Calibration curves for glycerol obtained at a flow rate of 0.5 mL/min with an injection loop of 250 µL were linear up to 10^{-3} mol/L with a detection limit of 10^{-6} mol/L and RSD of 1-2%. the working buffer was 0.1 M borate pH 8.5 containing 3 mmol/L ATP and Mg^{2+} (cofactors of GK).

A recovery study to ascertain the effect of the matrix (wine) on the biosensor performance was then carried out at different dilutions of the sample in the working buffer. The recovery values ranged between 90 and 103%, indicating the suitability of the procedure for wine (and must) analysis.

The FIA system was tested off-line for measurement of glycerol during fermentation with the same substrate (Trebbiano grapes must) but varying the initial pH and the temperature of the process. The results, reported in Figure 1 (triplicates), indicated that the influence of the initial pH whitin a 3.0-3.4 range was negligible for glycerol production. On the contrary, a difference in the temperature of the fermentation process (18 vs. 25 °C) influenced significatively both the kinetic of the production and the final content of glycerol.

These data demonstrated that because of its sensitivity and stability the developed biosensor is suitable for on-line analysis of glycerol during wine production.

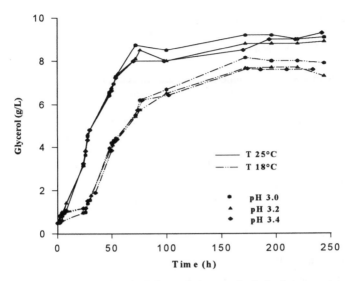

Figure 1 *Real-time monitoring of glycerol during alcoholic fermentation*

1.6 Lactulose

Lactulose is nowadays used as marker of heat treatment of milk by the International Dairy Federation (IDF)[8] and by the European Commission (EC)[9] to distinguish UHT milk from in container sterilised milk. The disaccharide (galactose + fructose), is not naturally present in raw milk, but it is formed during the heat treatment by isomerization of lactose. An upper threshold value of 60 mg per 100 ml of lactulose has been suggested in order to avoid an excessive heat damage in UHT milk.

We have realised a biosensor for lactulose determination by immobilisation the β-galactosidase enzyme on glass beads to have an enzyme reactor with a long lifetime, connected with a highly stable fructose biosensor. Finally the flow system was coupled with a microdialysis sampling technique, in order to obtain a constant and continuous recovery of lactulose from untreated milk samples and to allow a real continuous flow measurement of the lactulose at room temperature[10].

The enzymatic analysis of lactulose is based on the assumption that this compound is the only source of fructose in milk, and the latter is produced only by hydrolysis of lactulose, so the measurement of lactulose is directly correlated to the measurement of fructose.

The reactions involved are:

$$\text{Lactulose} + H_2O \xrightarrow{\quad \beta\text{-galactosidase} \quad} \text{Galactose} + \text{Fructose};$$

$$\text{Fructose} + 2\,[Fe(CN)_6]^{3-} \xrightarrow{\quad FDH \quad} \text{Ketofructose} + 2\,[Fe(CN)_6]^{4-} + 2H^+$$

$$2\,[Fe(CN)_6]^{4-} \xrightarrow[(+380\,mV)]{\quad Pt \quad} 2\,[Fe(CN)_6]^{3-} + 2e^-$$

Table 4. *Lactulose concentration (mg per 100 ml) in different milk samples comparison of the proposed method and the spectrophotometric kit by Boehringer. (Mean + SD, n=3; N.D.= Not detectable)*

Samples	Amperometric method (A)	Spectrophoto-metric method (B)	$E=\dfrac{A-B}{B}\%$
Pasteurised milk (whole)	4.2 (±0.2)	N.D.	-
Pasteurised milk (semiskimmed)	4.9 (±0.3)	N.D.	-
UHT milk (whole)	34.2 (±2.9)	36.2 (±2.3)	-5.5
UHT milk (semiskimmed)	30.8 (±1.8)	26.7 (±2.7)	-15.3
UHT milk (whole)	28.5 (±0.7)	29.7 (±2.0)	-4.0
UHT milk (semiskimmed)	18.8 (±1.5)	17.8 (±1.5)	5.6
Sterilised milk (whole)	114 (±9)	115 (±4)	-0.9
Sterilised milk (semiskimmed)	102 (±10)	124 (±4)	-17.7
Sterilised milk (skimmed)	99 (±8)	106 (±7)	-6.6

Fructose, produced by the hydrolysis of lactulose in the β-galactosidase reactor, is oxidised to ketofructose at the surface of the fructose biosensor, with the concomitant reduction of the ferricyanide mediator. This is reoxidised at the Pt electrode, giving a current proportional to the fructose, hence to the lactulose present in the medium.

Recovery tests, performed (at a flow rate of 0.1 mL/min) in different milk samples after a 1:1 dilution with McIlvaine buffer showed a percentage of recovery ranging from 95 to 105 % for pasteurised, UHT and in container sterilised milk.

Table 4 shows the comparison between the biosensor method and an enzymatic spectrophotometric reference procedure. Results are in good agreement, inter-day CV was 6-10%.This new procedure can provide a valid support in the analysis of lactulose in milk, as it is reliable and accurate; in addition the analysis time is significantly reduced. Also on-line measurements without pretreatment can be performed.

References

1. R. Antiochia and G. Palleschi, Anal. Lett., 1997, **30**(4), 683.
2. M.G. Lavagnini, D. Moscone, D. Compagnone, C. Cremisini and G. Palleschi, Talanta, 1993, **40**, 1301.
3. E. Marconi, G. Panfili, M.C. Messia, R. Cubadda, D. Compagnone and G. Palleschi, Anal. Lett., 1996, **29**, 1125.
4. R. Draisci, G. Volpe, L. Lucentini, A. Cecilia, R. Federico and G. Palleschi, Food Chem., 1998, **62,** 225.
5. M.C. Messia, D. Compagnone, M. Esti and G. Palleschi, Anal. Chem., 1996, **68**, 360.
6. A.C. Noble, G.F. Bursick, Am. J. Enol. Vitic., 1984, **35**, 110.
7. D. Compagnone, M. Esti, M.C. Messia, E. Peluso and G. Palleschi, Bios. & Bioel., 1998, **13**, 875.
8. International Dairy Federation, 1992, B-DOc. 222 IDF, Brussels.
9. EC Commission, 1992, Doc. VI/5726/92 Dairy Chemists' Group.
10. D. Moscone, R.A. Bernardo, E. Marconi, A. Amine and G. Palleschi, Analyst, 1999, **124**, 00.00.

FLOW-THROUGH ENZYME IMMUNOASSAY KITS FOR THE RAPID DETECTION OF THE MYCOTOXINS OCHRATOXIN A, T-2 TOXIN AND AFLATOXIN M_1 IN FOOD AND FEED

S. De Saeger, L. Sibanda and C. Van Peteghem

Laboratory of Food Analysis
Faculty of Pharmaceutical Sciences
University of Gent
Harelbekestraat 72
9000 Gent
Belgium

1 INTRODUCTION

Mycotoxins are toxic secondary metabolites produced by moulds during their growth on food and feed. Ingestion of such compounds can result in toxic syndromes in humans and animals. Ochratoxin A (OA) is a mycotoxin mainly produced by *Aspergillus ochraceus* and *Penicillium verrucosum*. OA contamination has been reported in cereals, coffee, and animal feed, as well as in pig tissues and pig blood. OA has carcinogenic, genotoxic, immunosuppressive, nephrotoxic, and teratogenic properties.[1] T-2 toxin (T2) is one of the trichothecene mycotoxins, formed by various species of *Fusarium*. T2 can be found in cereals. It is immunosuppressive, neurotoxic and dermatotoxic.[2-3] Aflatoxin M_1 (AFM) is a metabolite of aflatoxin B_1 (AFB) which is produced by *Aspergillus flavus* and *Aspergillus parasiticus*. AFM can be found in the milk of cows fed with AFB contaminated feed. Aflatoxins have hepatotoxic, carcinogenic, mutagenic, immuno-suppressive and teratogenic properties.[2,4]

Our objective was to develop kits for the detection of these mycotoxins in food and feed. The detection method was a flow-through enzyme immunoassay which resulted in rapid, visual and qualitative results.

2 MATERIALS AND METHODS

As flow-through device, a plastic snap-fit device from Trosley Equipment, Dover, Kent, England, was used. The bottom member of this device was filled with 100% cotton wool as absorbent material. A piece (2 by 2 cm) of Immunodyne ABC membrane (Pall Gelman Sciences, Champs-sur-Marne, France; pore size, 0.45 µm), coated with anti-mouse antibodies (Dako, Glostrup, Denmark), was placed on the absorbent material. The top member of the device was fitted on the bottom member. The membrane was placed on the absorbent pad such that the center of the membrane, where the antibody spot was located, was accessible through the aperture of the top member.[1,2,4]

Assay reagents were successively dropped on the membrane through the aperture of the top member with a micropipette.[1,2,4] First, 100 µl of a monoclonal anti-mycotoxin immunoglobulin solution (produced by L. Solti and I. Barna-Vetro, Institute for Animal

Sciences, Agricultural Biotechnology Center, Gödöllö, Hungary[5-7]) was dropped onto the membrane, followed by 300 µl of wash solution, 600 µl of mycotoxin standard solution or sample extract solution, 100 µl of mycotoxin-horseradish peroxidase (HRP) solution (prepared by L. Solti and I. Barna-Vetro, Institute for Animal Sciences, Agricultural Biotechnology Center, Gödöllö, Hungary[7-9]), and 600 µl of wash solution. We waited between the different steps until the liquid was absorbed by the absorbent pad. Finally, 50 µl of substrate-chromogen (H_2O_2 – tetramethylbenzidine) was dropped onto the membrane. After 1 min the dot colour intensity on the membrane was evaluated. Visual evaluation of the colour was done by comparing the dot colour intensity of the test membrane with that of the negative control, which showed the most intense blue colour because of the inverse relationship between toxin concentration and colour development. A portable colorimeter (Minolta Chroma Meter CR-321, Minolta Co., Ltd., Osaka) was used to quantify the colour of the dots on the membrane.[1-4]

Three flow-through enzyme immunoassay kits were produced for the detection of the mycotoxins OA, T2 and AFM in food and feed. All the reagents and materials necessary to perform the flow-through tests were included in the kit. The different components of the OA flow-through kit are listed in Table 1. The composition of the T2 and AFM kits was comparable with the OA kit.

3 RESULTS AND DISCUSSION

Three different kits were optimised. The first one included the OA test for the qualitative analysis of OA in wheat. Ground wheat samples (5 g) were mixed with 15 ml 80% methanol-water for 15 min. After filtration of the suspension through a Whatman no. 4 filter paper, the filtrate was diluted with wash solution by a factor of 3. This dilution was filtered through a 0.45 µm filter (Chromafil disposable filter, Macherey-Nagel, Düren, Germany). This filtrate was used in the flow-through immunoassay. There was no matrix interference.[1] With this extraction method and the aforementioned flow-through assay procedure a wheat sample with an OA content of 4 ng/g resulted in a complete colour suppression. When a blue coloured dot, even less intensive than the colour of the negative control, appeared on the membrane after 1 min, it indicated that the tested wheat sample contained OA at a level of less than 4 ng/g. When no coloured dot appeared on the membrane after 1 min, the OA level in the wheat sample was equal to or greater than 4 ng/g.[1] This 4 ng/g is the maximum permitted level for OA in cereals proposed by the European Commission.[10]

The second kit contained a test for the qualitative determination of T2 in wheat, barley, rye and maize. The same simple and rapid extraction method as for the OA test was used. There was no matrix interference. The detection limit and cut-off value for this test was 50 ng/g T2 in cereals. This means that when a blue coloured dot appeared on the membrane after 1 min, the T2 concentration in the sample was less than 50 ng/g. When there was no colour development, it indicated that T2 was present in the sample at a level greater than or equal to 50 ng/g. There is no existing European maximum permitted level for this mycotoxin.

The third kit included a flow-through test for the detection of AFM in milk. The sample preparation was done with immunoaffinity columns (Aflaprep M; Rhône-Diagnostics Technologies, Glasgow, UK).[4] The cut-off value of this flow-through test was 0.01 ng/ml AFM in milk. In the Commission Regulation (EC) No 1525/98 a

Table 1 *Composition of the OA kit*

REAGENT	AMOUNT	PACKAGING
flow-through devices filled with cotton wool and anti-mouse coated membranes	10 pieces	plastic vacuum-sealed bag
wash solution	2 x 15 ml	plastic bottle
assay buffer	6 ml	plastic bottle
negative control	4 ml	plastic bottle
positive control	4 ml	brown glass vial
colour reagent A	500 µl	brown glass vial
colour reagent B	500 µl	brown glass vial
anti-OA immunoglobulin (undiluted)	10 µl	glass vial
OA-horseradish peroxidase (undiluted)	15 µl	brown glass vial
Whatman filters	6 pieces	plastic bag
plastic 3 ml syringes	6 pieces	-
0.45 µm filters	6 pieces	-

maximum level for AFM in milk of 0.05 ng/ml has been set. However, by adjusting the antibody and enzyme conjugate dilutions, it is possible to modify the cut-off value such that it is in compliance with the Commission Regulation.

The described mycotoxin kits are valuable tools for screening of food and feed. With one kit six samples can be analysed. Two controls and three samples can be analysed together and qualitative results are obtained in less than 30 min (time for extraction not included). A quick discrimination between positive and negative samples can be made. Positive results must be confirmed by methods other than immunoassays. To test the applicability of the mycotoxin kits in different laboratories an interlaboratory study has been started.

4 ACKNOWLEDGMENTS

This work was financially supported by the Ministry of the Flemish Community, Economy Administration (Cooperation Flanders and Central and Eastern Europe. Contract No 508/97-06-20/02-03-04-16).

References

1. S. De Saeger and C. Van Peteghem, *J. Food Prot.*, 1999, **62**, 65.
2. S. De Saeger and C. Van Peteghem, European Patent Application EP 0 893 690 A1.
3. S. De Saeger and C. Van Peteghem, *Appl. Environ. Microbiol.*, 1996, **62**, 1880.
4. L. Sibanda, S. De Saeger and C. Van Peteghem, *Int. J. Food Microbiol.* (in press).
5. A. Gyöngyösi-Horvath, I. Barna-Vetro and L. Solti, *Lett. Appl. Microbiol.*, 1996, **22**, 103.
6. A. Gyöngyösi, I. Barna-Vetro and L. Solti, 'ECB6: Proceedings of the 6th European Congress on Biotechnology, Florence, Italy, June 13-17, 1993', L. Alberghina, L. Frontali and P. Sensi (ed.), Elsevier Science B.V., Amsterdam, 1994, p. 709.

7. I. Barna-Vetro, L. Solti, E. Szabo and A. Gyöngyösi-Horvath, 'Proceedings of the International Symposium Environmental Biotechnology, April 21-23, 1997, Oostende, Belgium', 1997, p. 73.
8. I. Barna-Vetro, A. Gyöngyösi and L. Solti, *Appl. Environ. Microbiol.*, 1994, **60**, 729.
9. I. Barna-Vetro, L. Solti, J. Teren, A. Gyöngyösi, E. Szabo and A. Wölfling, *J. Agric. Food Chem.*, 1996, **44**, 4071.
10. J. E. Smith, C. W. Lewis, J. G. Anderson and G. L. Solomons, 'Mycotoxins in human nutrition and health', Science Research Development, European Commission, Brussels, 1994, p. 74.

EVALUATION OF NOVEL *IN VITRO* ASSAYS FOR THE DETECTION OF BOTULINUM TOXINS IN FOODS

M. Wictome, K. Newton, K. Jameson, P. Dunnigan [†], S. Clarke [*], S. Wright [*], J. Gaze [*], A. Tauk [*], K. A. Foster and C. C. Shone.

Centre for Applied Microbiology and Research, Porton Down, Salisbury, Wiltshire, SP4 OJG, U.K., [†] Rhône Diagnostics Technologies, West of Scotland Science Park, Glasgow, G20 OSP, U.K., [*] Campden and Chorleywood Food Research Association, Chipping Campden, GL55 6LD, U. K.

1 INTRODUCTION

Various strains of the bacterium *C. botulinum* produce a family of seven structurally related but antigenically different protein neurotoxins (types A to G) which cause the syndrome botulism.[1] Symptoms are presented as widespread flaccid paralysis which often results in death. Much effort has been imparted by the food industry to ensure that food treatment processes prevent the growth and toxin production of *C. botulinum* and there is a need for rapid, sensitive and specific assays for these toxins.

At present the only method of confidence in the detection of the toxins is the acute toxicity test performed in mice.[2] Although this test is exquisitely sensitive, with a detection limit of 1 mouse 50% lethal dose [MLD_{50}] being equivalent to 10-20 pg of neurotoxin/ml, it has a number of drawbacks: it is expensive to perform, requires a large number of animals and is not specific for the neurotoxin unless neutralisation tests using a specific antiserum are carried out in parallel. In addition the test takes up to 4 days to complete. The increasing resistance to such animal tests has also required the development of alternative rapid *in vitro* assays with the sensitivity and reliability of the mouse bioassay.

Over the past five years significant progress has been made in deciphering the mode of action of the botulinum neurotoxins (BoNT). The toxins have been demonstrated to act at the cellular level as highly specific zinc endoproteases cleaving various isoforms of three small proteins controlling the docking of the synaptic vesicles with the synaptic membrane. Botulinum neurotoxins A and E cleave specifically the 25 kDa synaptosomal associated protein (SNAP-25). Botulinum neurotoxin C cleaves the membrane protein syntaxin and SNAP-25. Botulinum neurotoxin types B, D, F and G act on a different intracellular target, vesicle-associated membrane protein (VAMP) or synaptobrevin.[3] Whilst the endopeptidase activity of the neurotoxin has been utilised in assay formats such assays do not have the sensitivity required to act as replacements for the mouse bioassay nor are they suitable for the detection of BoNT in food samples.[4] Here we describe an assay format, based on the endoprotease activity of the neurotoxin, with a sensitivity that exceeds that of the mouse bioassay and which is sufficiently robust to allow the detection of neurotoxin in a range of food-stuffs.

2 MATERIALS AND METHODS

Preparation of type B and E neurotoxin. BoNT/B (Okra) or BoNT/E (NCTC 8550) was cultured in Cooked Meat Carbohydrate Medium (Oxoid) for 48 hr at 37°C . The level of neurotoxin was assessed by sandwich ELISA calibrated against purified neurotoxin. The culture supernatants were trypsin activated and diluted to 1ng/ml (neurotoxin) in 50 mM Hepes, 150 mM NaCl, 1 mg/ml bovine serum albumin, pH 7.4 and stored at -70°C. The toxicity of this toxin standard was assessed using the mouse bioassay as described previously.[2] *Synthesis of type B peptide substrate.* A peptide substrate representing residues 60 to 94 of human VAMP isoform 1 [VAMP (60-94)] was synthesized, purified and characterised as described.[4] An C-terminal cysteine was added to the peptide which was post-synthetically modified to incorporate a biotin moiety. *Synthesis of the type E substrate.* Recombinant mouse SNAP-25 was generated as a fusion protein with glutathione-*S*-transferase using a pGEX expression system (Pharmacia) and purified on glutathione affinity column according to the manufacturers instructions. Full-length SNAP-25 was generated by cleavage of the fusion protein with thrombin. *Production of antibodies to cleaved susbstrates.* Antisera was raised against the peptides FESSAAKC and CQNRQIDR which represent the C-terminal and N-terminal side of the cleavage site on VAMP and SNAP-25 by BoNT/B and BoNT/E respectively.[4] *Preparation of immunoaffinity columns.* Immunoaffinity columns for the extraction of the neurotoxin were prepared using cyanogen bromide activated Sepharose 4B (Pharmacia-LKB) following the manufacturers instructions. 1 ml of gel slurry was then added to a disposable plastic column (65 mm x 10 mm, Rhône Diagnostics Technologies) giving approximately 100 µl of packed gel, containing 10 µg of immobilised Ab per column. (BoNT/B assay: 5µg 5BB/21.3, 5µg 5BB/9.3; BoNT/E assay: 10 µg polyclonal Ab) The columns were stored at +4°C. *Preparation of Food Extracts.* Food samples were shaken vigorously with an equal volume of gelatin-phosphate buffer and stored at +4°C for 18 hrs. The sample was then centrifuged at 13,000 g for 20 min at 4°C, after which the supernatant was removed and spiked with $1MLD_{50}$ ml $^{-1}$ of neurotoxin. Extract was then assayed by both the endopeptidase assay and mouse bioassay. *BoNT/B Assay.* Storage buffer was drained from the immunoaffinity columns and 2ml samples were added. The columns were then sealed and shaken horizontally at 37°C for 15 min. After which the columns were washed three times with 2.5 ml of 50 mM Hepes, 20 µM $ZnCl_2$, pH 7.4 (HZ buffer) and the column drained. 100 µl of 25 µM peptide substrate was added in HZ buffer containing 10 mM dithiothreitol (HZ-DDT). The columns were then shaken at 37 °C for 2 hr in an upright position. 400 µl of phosphate buffered saline containing 0.1% Tween-20 (PBS-Tw) was then added to each column, mixed and 100 µl samples transferred to a streptavidin coated microtitre plate. The plate was then shaken for 5 min at 37 °C, after which unbound material was removed by washing. Antibody specific to the cleaved peptide was then added (1.2 µg/ml in PBS-Tw containing 5% foetal calf serum) for 1 hr at 37 °C. Unbound material was removed by washing, after which rabbit anti-guinea pig immunoglobulin horseradish peroxidase conjugate was added for 1hr at 37 °C. After washing tetramethylbenzidine substrate solution was added. *BoNT/E Assay.* The format of the assay was as described for the BoNT/B assay with the following exceptions. After incubation of the sample the columns were washed with 2.5

ml of HZ buffer followed by 2 x 2.5 ml of deionised water. 0.5 ml of 5 mM glycine, pH 2.7, was then added and the columns shaken at 37 °C for 10 min, after which the columns were drained and 50 µl of x 10 strength HZ-DTT buffer was added to the eluate . 100 µl samples were then added to a microtitre coated with SNAP-25 and incubated at 37 °C for 2 hr. Cleaved peptide fragments generated were detected with 400 ng/ml of antibody incubated at 37 °C for 1 hr.

3   RESULTS AND DISCUSSION

In the present study a novel, rapid, *in vitro* assay has been developed for the detection of neurotoxin from type E and proteolytic type B strains present in food products. An assay system for the detection of type B neurotoxin is depicted in Figure 1. After immunoaffinity capture of the toxin, using mAb specific for proteolytic type B neurotoxin, unbound material is removed by washing and a biotinylated peptide substrate is added which the immobilised toxin cleaves. The peptide fragments are then captured on a streptavidin coated microtitre plate and the cleaved peptide fragments detected using an antibody which is specific to the newly exposed N-terminus of the peptide substrate. The detection limit for the assay for was found to be approximately 0.5 MLD_{50}/ml, employing a cut-off of 0.5 absorbance units above background, data not shown. The type E assay has a similar format and differs in that the toxin is eluted from the column prior to substrate cleavage.

Unlike previous solid-phase based assays the assay format described can be used in a range of test media, as the non-bound material that may interfere with the endopeptidase activity of the toxin is removed by washing prior to addition of the peptide substrate.

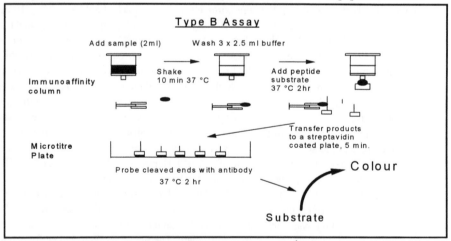

Figure 1 *BoNT/B assay* ▭▭ *Immobilised mAb* ⊔ *Biotinylated Substrate*
             Streptavidin ● *Neurotoxin*

Table 1 demonstrates the ability of the assay to detect 1 MLD_{50}/ml in spiked pate, cod and cheese extracts for the type B and E assay. The assay format described only detects active and not denatured toxin and in this respect the assay more closely reflects the mouse bioassay than a conventional ELISA. To date no false positive results have been encountered, the result of the combined specificity of the capture Abs and that of the endopeptidase activity of the toxin.

Table 1 *Detection of 1 MLD$_{50}$/ml of BoNT/B and BoNT/E in foods. OD 450 nm above blank of <0.1. Figures in parenthesis are % CV and n respectively.na: not assessed.*

Food	BoNT/B	BoNT/E
Pate	2.26 (25.5, 50)	1.35 (27.1, 6)
Cheese	1.83 (39.4, 49)	0.30 (58.9, 6)
Cod	2.58 (11.2, 54)	0.70 (11.7, 6)
Beef	1.22 (64.8, 36)	na
Yogurt	2.04 (11.3, 6)	na

Table 2 *Comparison of the ability of the BoNT/B endoprotease assay (EA) and the mouse lethality test (MT) to detect crude neurotoxin present in spiked food samples.*

BoNT/B (pg/ml)	Pate EA	MT	Sausage EA	MT	Buffer EA	MT
50	2.78	+	2.02	+	2.57	+
25	2.80	+	1.90	+	2.54	+
12.5	2.81	+	2.09	+	2.69	+
6.3	2.15	+	1.82	-	1.80	+
3.1	1.24	-	1.19	-	1.09	-
1.5	0.78	-	1.05	-	1.02	-
0	0.02	-	0.004	-	0.04	-

Table 2 shows the sensitivity of the BoNT/B assay compared with the mouse test. At lower concentration of toxin the sample is non-toxic in the *in vivo* assay but gives a positive result in the endopeptidase assay. The sensitivity of the assay allows a visual reading and it is predicted that such assays will be used to give a yes/no answer for the presence or absence of toxin in a sample within 5-6 hr. A similar approach has also resulted in the development of assay formats for the detection of neurotoxin from serotypes A and F. Whilst extensive validation of the assay formats is required before they will replace the mouse bioassay it is hoped the assays will greatly reduce the number of mice used in the detection of botulinum neurotoxins.

References

1. C. L. Hatheway, Curr. *Top. Microbiol. Immunol.*, 1995, **195**, 55.
2. L. J. Reed and H. Muench. *Am. J. Hyg.*, 1938, **27**, 493.
3. C. Monteccuco, G. Schiavo, V.Tugnoli and D.de Grandis, *Mol. Med.Today*, 1996, **2**, 418.
4. B. Hallis, B. A. F. James and C. C. Shone, *J. Clin. Microbiol.*, 1996, **34**, 1934.

Acknowledgements:The financial assistance of the Ministry of Agriculture, Fisheries and Food is gratefully acknowledged.

DNA BASED METHODS FOR MEAT AUTHENTICATION

R. G. Bardsley and A. K. Lockley

Division of Nutritional Biochemistry, School of Biological Sciences, University of Nottingham, Sutton Bonington Campus, Loughborough LE12 5RD

1 INTRODUCTION

Consumer groups and the general public are likely to become increasingly insistent that food products should not only be free from contamination by microorganisms and chemical residues, but should contain only authentic ingredients from acceptable sources. Recent concern about BSE has drawn attention to the incorporation of bovine tissues in animal food products, while for foods of plant origin, the unannounced introduction of genetically-modified crops has led to widespread and even violent reaction. The difficulties associated with authentication of meat components are obviously greatest in processed or composite food mixtures where the origins of the components is not readily apparent. For these products there is likely to be an increasing demand for more accurate and rapid analysis, that may in the future extend even to the level of animal breeds or plant varieties as consumers become more sophisticated.

Laboratories involved in analysis of animal products have formerly relied largely on immunochemical procedures,[1] which usually necessitate the production of specific antisera to each target organism, with unwanted cross-reactivity being an ongoing problem which places limits on sensitivity. Batch-to-batch variability of antisera must continually be guarded against, and furthermore it is often necessary to create an additional or parallel series of antisera against a target tissue which has been processed in various ways. Processing can lead to partial or complete loss of the epitopes normally present in native protein structures. Until about ten years ago, little attention appears to have been paid to the residual nucleic acid in animal materials which potentially retains considerable information about species of origin. This neglect reflects the well-known technical difficulties associated with nucleic acid analysis, namely the great instability of ribonucleic acid and the demanding methodologies of molecular biology, including electrophoresis, blotting, cloning and use of hazardous ^{32}P-labelled probes. The advent of the polymerase chain reaction[2] with its capacity for rapid amplification of minute quantities of residual DNA fragments, appears to offer unlimited potential for the food analyst and is being extensively researched for this purpose.

In this overview, we have attempted a brief survey of the development of methods for meat authentication via DNA, although it is not intended for this to be a comprehensive review. We shall divide the topic broadly into three sections; the early hybridisation approach which in many ways paralleled conventional immunochemical methods; current experimentation which is almost completely centred around the polymerase chain reaction; a number of new developments which are on the horizon, driven for the most part by the clinical diagnostics industry.

2 MEAT SPECIATION USING HYBRIDISATION PROBES

2.1 Dot blots

In many early methods, DNA fragments recovered from the tissues to be analysed were spotted onto a nitrocellulose or nylon membrane and irreversibly bound by UV crosslinking. Genomic DNA fragments from a particular meat species, and therefore presumed to contain sequences unique to that species, were labelled for use as probes, using radioisotopes or various haptens which could be detected with appropriate primary antibodies and conjugated secondary antibodies for colorimetric visualisation. Under denaturing conditions, primarily high temperature at 95°C, the strands in target and probe DNA separate and reanneal to incorporate labelled probe into the immobilised complementary sequence if present. Using probes prepared from pig, chicken, goat, sheep and cattle DNA, the degree of cross-reactivity between species could be assessed.[3,4] By adjusting annealing temperature, the presence of pork at approximately 0.5% in meat mixtures could be detected, although there was less discrimination using probes for one ruminant species against another, due to the close genetic relationship. Interestingly, hybridisation of specific DNA sequences appeared to survive heat treatment of raw meat, typically at 100-120°C for an hour, which simulates some aspects of processing. This early approach did not require any knowledge of unique sequences in the genome of a particular species, although subsequently the use of other types of probes, including synthetic oligonucleotides[5] or cloned cDNAs based on species-specific sequences such as a tandem repeat would become available.[6]

2.2 Restriction fragment length polymorphisms (RFLP)

The use of restriction enzyme digestion adds a second level of discriminatory capacity to the use of DNA probes. Such enzymes are absolutely specific for sequences as short as four to six nucleotides and high molecular weight DNA can readily be digested to completion, generating reproducible fragmentation patterns after agarose gel electrophoresis. The DNA can be transferred to nitrocellulose or nylon membranes by Southern blotting, which can then be hybridised as above to labelled cDNA or oligonucleotide probes. The resulting patterns reflect sequence variability at one or several restriction sites in the vicinity of a complementary target sequence, rather than merely the presence or absence of the hybridising sequence itself. When using probes based on repetitive mini- or micro-satellite sequences, characteristic individual-specific patterns were discovered[7] which formed the basis of classical DNA fingerprinting used in human paternity testing and forensic science. However, individual-specific fingerprints are generally considered too specific for speciation purposes, although Blackett and Keim[8] were able to obtain species-specific profiles for elk, deer and antelope using probes based on repetitive sequences.

The relatively high abundance of mitochondrial DNA compared to genomic DNA has made it an attractive target for RFLP analysis, and probes based on the mitochondrial genome have been able to discriminate approximately twenty species of large mammal.[9]

2.3 RFLP analysis of the actin multigene family

In our laboratory, the well known high conservation of the amino acid and nucleotide sequence of the ubiquitous cytoskeletal and myofibrillar protein actin prompted us to examine the actin gene(s) in meat animal species by RFLP analysis. Using a murine actin cDNA construct as a probe, species-specific 'fingerprints' comprised of 10-20 bands were obtained for a number of meat species (Figure 1).[10] The patterns survived heat treatment at 120°C and were essentially independent of breed. The sizes of the various bands suggested that they were derived from different members of the actin multigene family, rather than from fragmentation of a single actin gene. Since the actin genes are known not to be grouped on any one chromosome in mammals, these gene sequences can be regarded as 'large repeats' occurring infrequently and with wide distribution throughout the mammalian genome. Although the actin 'fingerprints' were highly reproducible and informative, their value in routine analysis of meat mixtures would be reduced because of the complexity of the combined patterns. As with many hybridisation techniques, a number of factors influence annealing efficiency and it is often difficult to control the relative intensities of the various bands, especially in mixtures. Accordingly, any test would have to have a range of internal DNA standards, and the stringency of hybridisation would have to be carefully controlled.

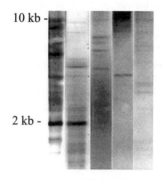

Figure 1 *RFLP analysis based on the actin multigene family in meat species (L-R: sheep, pig, chicken, horse, cattle).*

3 THE POLYMERASE CHAIN REACTION AND MEAT SPECIATION

3.1 Mechanism

Hybridisation techniques such as RFLP tend to require relatively large quantities (~10 μg) of good quality high molecular weight DNA, which is not easily recoverable from small amounts of fresh meat or processed mixtures. Fortunately, in the last ten to fifteen years a number of techniques have been perfected for the amplification of minute quantities of a DNA target sequence, even the two copies (or alleles) of a gene sequence normally present in a single diploid cell. The most widely deployed method is undoubtedly the polymerase chain reaction (PCR)[2], based on the use of complementary synthetic oligonucleotides 20-30 nucleotides long to prime the copying of both 'sense' and 'antisense' strands of a region of DNA duplex simultaneously. By use of a thermostable DNA polymerase, the original duplex can be repeatedly melted, reannealed to primers and recopied, as indeed can the products of each round of amplification.   Thus an approximately exponential reproduction of a specified region of a DNA sequence can be continued during about 30 cycles of copying, to the point where the DNA can readily be visualised on agarose gel electrophoresis.

3.2 PCR and speciation

3.2.1 Introduction. PCR may be used in a number of ways for authentication purposes. Ideally, the design of the primers, the ionic composition of the buffers and the thermal cycling parameters should be adjusted so that a visualisable product of the predicted size is only seen when a target sequence is present in the material under test. However, the failure to detect a product might be the result of an ineffective polymerase, or a missing or degraded primer or other experimental artefact, so that a positive control using a standard DNA template should always be included to guard against procedural errors. In DNA from a mixed meat product, the presence of a band of correct size might result from amplification of highly similar DNA sequence from an unknown species, constituting a false positive. The identity of closely related products can best be confirmed by full sequencing, a procedure which has been termed 'forensically informative nucleotide sequencing' (FINS). The cytochrome b gene (*cytb*) in mitochondrial DNA (mtDNA) has been a favourite target sequence for many FINS studies, for example in tuna speciation.[11] Although DNA sequencing is becoming increasingly rapid with the advent of automated equipment, its use as a routine tool would not be applicable for many smaller laboratories, so that simpler PCR-based tests will continue to be sought. Fortunately, sequence variability in PCR amplicons of the same size can be detected by a number of post PCR steps.

3.2.2 CAPS and SSCP. A number of analysts have made use of the strategy in which primers capable of annealing to DNA templates from different species generate a PCR product or amplicon of constant size, but with variable sequence. The amplicon is then subjected to restriction enzyme digestion and the fragment sizes produced then become indicative of species present, a technique known as 'cleavage of amplification products' (CAPS). For example, amplification of mtDNA *cytb* sequences allowed discrimination of tuna[12,13] or deer[14] species by the CAPS technique. Because the CAPS approach requires one or more post-PCR enzymic steps, other methods for the demonstration of amplicon sequence variability have been developed. One of these exploits possible differences in secondary structure which can affect nucleic acid mobility on polyacrylamide gel electrophoresis. Although technically considered difficult, once optimised, the 'single strand conformational polymorphism'(SSCP) method has proved useful for rapid detection of sequence variability, for example in tuna speciation.[15,16]

3.2.3 Speciation by amplicon size. In FINS, CAPS and SSCP, common or universal primers capable of annealing to a target sequence such as the *cytb* gene in many species generate a PCR product of the same size, which then requires further characterisation.   In other approaches, one universal forward primer is used in combination with a second species-specific reverse primer. By careful primer positioning, the size of the PCR amplicon can then be made indicative of the various species present. Again targetting the mtDNA *cytb* gene, Matsunaga and coworkers[17] have designed a multiplex reaction using one universal primer and several species-specific primers which is capable of generating a 'ladder' of PCR products, with each rung on the ladder being indicative of a species present.

3.2.4 PCR and the actin multigene family. In an earlier section, the unusual features of the mammalian actin gene family, namely its multiple copies in most species and highly conserved coding sequences, were shown to be useful for obtaining species-specific information by RFLP analysis. This conservation of coding sequences makes it possible to design universal PCR primers potentially capable of interacting with all actin genes, either within or between species. When such primers were designed to straddle the site of known introns, a series of PCR products of different sizes was generated with all the meat species tested.[18] On agarose gel electrophoresis, 'fingerprints' characteristic of a species were seen which were reproducible and survived heat treatment at 120°C (Figure 2). The origins of the different bands is the length of intronic sequence between the highly conserved exons common to each gene copy. Although the fingerprints were generally informative for single species, they would be unsuitable for analysis of meat mixtures because of the complexity of the combined patterns. A better strategy for this purpose would be to design primer pairs which give a single unique band for each species. This was achieved by retaining one common primer in an exon and selecting a second primer in the adjacent intron where less conservation of sequence was expected. This procedure allowed a PCR reaction to occur which was highly selective for chicken DNA, with cross reaction only with turkey DNA and none of the other avian species tested.[19] Subsequently, primers based on intron sequences in chicken and turkey genes have been designed by our group which are completely specific to their target species, giving bands of characteristic size which can be used to detect less than 1% chicken in the presence of 99% turkey and *vice versa* (unpublished work).

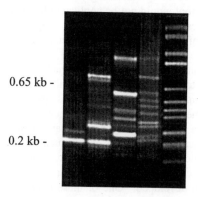

0.65 kb -

0.2 kb -

Figure 2 *PCR of the actin multigene family in meat species (L–R: horse, sheep, chicken, cattle, markers)*

3.2.5 ARMS and MS-PCR. The design of primers to generate amplicons whose size is indicative of species can be approached using a further strategy, based on the fact that the 3' terminal nucleotide of a primer is critical for its extensibility. Since there must be exact complementation between primer and template in this position, primers can in principle be designed which differ only in their 3' terminal base and are therefore capable of detecting a single base difference between two otherwise identical templates. This technique has been termed 'amplification of refractory mutations' (ARMS),[20] or in a modified form 'mutagenically-separated' or MS-PCR.[21] To detect a mutation in the porcine ryanodine receptor gene (*RyR1*), Lockley and coworkers[22] designed two forward primers whose sequence was essentially identical for some twenty nucleotides of their length, so that they would both anneal to the same target sequence; however, one primer carried an additional 20 nucleotides at its 5' terminus, whose sequence is not critical, while the 3' terminal nucleotide was modified to be complementary to the mutant rather than normal genotype. A single reverse primer was used. In this way, a 20 bp differential in size of amplicons, established by agarose gel electrophoresis, was sufficient to be indicative of genotype at this locus. A similar approach has been used in speciation of tuna, where a single base difference in bonito and bluefin mtDNA *cytb* sequences was sufficient to allow genotyping (Figure 3, unpublished work).

20 bp differential ⊂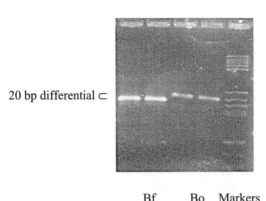

Bf   Bo   Markers

Figure 3 *Use of MS PCR to authenticate bonito (Bo) and bluefin (Bf) tuna.*

3.2.6 Current status of PCR in meat speciation. It will be apparent from the above discussion that PCR is a versatile technique that in principle can be applied to meat speciation in a variety of ways. Although the theoretical basis for testing for most meat species may have been demonstrated experimentally, extensive development, optimisation and trouble shooting will still necessary so that one or more of these test strategies may ultimately prove robust enough to be used in food analytical laboratories throughout the world with relatively simple equipment and basic skills. The almost total reliance on some form of gel electrophoresis may be seen by many as a barrier to widespread deployment, and various strategies have been tried to link the now 'classical' PCR reaction in microcentrifuge tubes to kit versions giving a more direct readout. Equally important is the fact that food analysts will increasingly require quantitative information about whether the levels of 'adulterant' species in composite foods is above that permitted by legislation. The standard PCR reaction is at best semi-quantitative, and an extensive series of controls is often required to obtain measurement which are sufficiently reliable to reflect the copy number or abundance of a particular target sequence in a meat or fish product.

4 POTENTIAL DEVELOPMENTS IN MEAT AUTHENTICATION

4.1 Avoidance of gel electrophoresis

The power of the PCR technique or its close variants is likely to ensure that it remains central to most future strategies of diagnosis based on nucleic acid sequences. This will also be true for the clinical diagnostics field, from where, because of the greater financial resources available and the intense pressure for the accuracy of diagnosis, the most rapid, reliable and robust formats may be expected to emerge. The current widespread use of gel electrophoresis undoubtedly slows down DNA tests and reduces the scope for automation.

Accordingly, a number of groups have attempted to develop solid phase or immobilised PCR, or PCR linked to fluorescence signals which can be quantified in end-point or real-time assays. A few comments outlining our own experiences and observations are offered as an introduction to this rapidly developing field, although no attempt has been made at a comprehensive review.

4.2 Immobilised reactions

Oligonucleotides can be immobilised to various solid supports by their 5' termini, in the hope that the free portion of the strand can still hybridise to complementary sequences, or even serve as an extensible primer for DNA polymerases in general and PCR in particular. The 5' terminus of an oligonucleotide can be tagged with a polyT sequence which can be irreversibly attached to a nitrocellulose-based membrane by UV cross- linking.   Such immobilised probes can be used to hybridise to and capture mRNAs, cDNAs or PCR products tagged by haptens or fluorophores, to be revealed immunochemically or under an appropriate lamp.   In this way non-electrophoretic tests for various HLA loci were developed.[23]   Our group attempted to use this method of primer anchoring to design a semi-solid phase reaction to detect the single base mutation in the *RyR1* gene (section 3.2.5) by MS PCR.[24]   Two essentially identical 'forward' primers differing in their 3' terminal nucleotide were 5' tailed with an 80 residue polyT sequence and crosslinked as two separate spots on triangular 'dipsticks' cut from nylon membranes.

Figure 4 *Immobilised MS PCR to distinguish (from L-R) normal, heterozygote and mutant RyR1 genotypes in pigs.*

The triangles were placed in standard microcentrifuge tubes and the PCR reaction carried out in a total volume of 100μl. The reverse primer was free in solution, and digoxigenin-dUTP (DIG) was supplied in the anticipation that it would be incorporated into the primer at the upper, lower or both spots of primer in the presence of DNA from any of the three genotypes, to be subsequently visualised immunochemically. The method successfully identified the normal, mutant and heterozygous genotypes as appropriate combinations of coloured spots (Figure 4), and may have potential in the future as a simple method for meat authentication.

The polyT crosslinking step is relatively non-specific and other methods are available for oligonucleotide anchoring, including conjugation of a terminal 5' phosphate group to a modified nylon base (Nucleolink, Nunc) by a phosphoramidate bond.[25] Although in an early stage of development, immobilisation of MS-PCR primers for tuna speciation (section 3.2.5) has been coupled to microtitre plate wells by this means and used to incorporate DIG- or fluorescein-tagged nucleotides into the extensible primers specific for bluefin or bonito DNA templates (unpublished work).

In the clinical diagnostics area, the principle of using immobilised capture probes has been refined to the point that many thousands of oligonucleotides can be synthesised on chips by a photolithographic procedure.[26] These serve to capture fluorescently-labelled nucleic acids at appropriate points on the array, which is then read by an addressable laser. From the resulting patterns of fluorescence after computer analysis, an accurate diagnosis of RNA variants or nucleotide sequence can be obtained. Other array technologies are being developed in which several hundred wells contain immobilised primers which will extend by a few nucleotides in the presence of an appropriate polymerase and the correct DNA template. The mass and therefore the identity of the extended primer and incorporated nucleotides is obtained by an addressable laser in conjunction with the MALDI TOF technique.[27]

The instrumentation and chips for the array technologies being developed for clinical diagnosis are currently expensive but may eventually find a role in food authentication. They are primarily designed to detect sequence variability, mutations and genotype and, in principle, many of the basic PCR strategies discussed earlier in this article (section 3) may be directly adaptable to different types of chip technology platforms. However, the element of quantification which is especially important for food analysis may not be appropriate for this particular approach.

4.3 Fluorescence and PCR

A number of fluorophores have now been developed for attachment to the 5' terminus of PCR oligonucleotide primers. In general these should have little effect on the annealing of the primer to its target sequence, and none at all on the extensibility at the 3' terminus, giving great scope for future developments. By use of primers tagged in this way, the ensuing PCR amplicons will fluoresce at an appropriate wavelength and can be detected quantitatively as they pass down an electrophoresis slab or capillary gel, or are read in a fluorimeter without an electrophoretic step. These procedures could readily be applied to most of the PCR methodologies for meat speciation described in earlier sections. In the absence of a gel step, however, no indication is given that the amplicon is of the correct size and removal of unincorporated fluorophore must be undertaken to avoid the problem of high backgrounds.

To circumvent the problem of high background fluorescence, a number of promising approaches will be mentioned. The first of these introduces the concept of the 'molecular beacon', a customised oligonucleotide sequences designed to hybridise to the product of a PCR reaction.[28] In addition the beacon is designed with a sequence at its 5' terminus which is complementary to a sequence at the 3' terminus, so that in the absence of target sequence the beacon forms a self-annealing stem-loop structure. Also at 5' and 3' termini, respectively, is attached a fluorophore and a corresponding quencher molecule, comprising a 'fluorescence resonance transfer energy transfer' (FRET) system. Thus in the stem-loop no signal is generated until the beacon interacts with a complementary PCR amplicon and unfolds, relieving the quenching effect. If a beacon is included in the PCR mix, no signal is generated until sufficient PCR product accumulates. Instrumentation is being developed to monitor the development of such fluorescence in real time,[29] or alternatively the endpoint fluorescence can be quantified by a standard fluorimeter. A related version of the molecular beacon technique relies on a stem-loop containing the FRET couple which is an integral part of the 5' terminus of a PCR primer itself.[30] Finally, in other versions, there is cleavage of a beacon-like FRET molecule attached to template DNA as each annealing and extension cycle of PCR takes place.[31]

The introduction of fluorescence into PCR quantification is likely to be a most important step forward for food authentication laboratories with their requirement for establishing legal or illegal levels of permitted or other adulterants. Most of the basic PCR strategies described in earlier sections should be adaptable to the use of the growing range of fluorescent PCR methods, using different fluorophore 'colours' suitable for multiplex PCR, so that rapid developments can be expected in this area.

5 CONCLUSIONS

The development of DNA-based tests for food authentication began about ten years ago and is beginning to be reliable enough for routine analysis. A number of commercial organisations offer tests, mostly using conventional PCR followed by gel electrophoresis. Methods will continue to be sought which can speed up the testing procedures, for example by the introduction of automation, and it is possible that some of the advanced procedures being developed for clinical diagnosis will prove suitable for food analytical laboratories. Although not discussed in this article, the rapid recovery of PCR-compatible DNA from complex mixtures or highly processed food products is not a trivial matter, and improvements are certainly needed here in order to exploit the full potential of the range of DNA-based tests currently under evaluation.

The authors are pleased to acknowledge the University of Nottingham-Allied Lyons Research Fund, the Ministry of Agriculture, Fisheries and Food, the Meat and Livestock Commission and the Department of Trade and Industry for financial assistance, and Drs. Karen Fairbrother, Andrew Hopwood, Stephen Franklin, John Dooley and Tim Parr for ideas and input over a number of years.

References

1. R. L. S. Patterson and S. J. Jones, *Analyst*, 1990, **115**, 501.
2. K. B. Mullis and F. A. Faloona, *Meth. Enzymol.*, 1987, **155**, 335.
3. K. Chikuni, K. Ozutsumi, T. Koishikawa and S. Kato, *Meat Sci.*, 1990, **27**, 119.
4. K. F. Ebbehøf and P. D. Thomsen, *Meat Sci.*, 1991, **30**, 221.
5. G. Nelson, P. F. Hamlyn, L. Holden and B. J. McCarthy, *Textile Res. J.*, 1992, **62**, 590.
6. J. B. Buntjer, J. A. Lenstra and N. Haagsma, *Zeitscrift für Lebensmittel-Untersuchung und Forschung*, 1995, **201**, 577.
7. A. J. Jeffreys, V. Wilson and S.L. Thein, *Nature*, 1985, **316**, 76.
8. R. S. Blackett and P. Keim, *J. Forensic Sci.*, 1992, **37**, 590.
9. M. A. Cronin, D. A. Palmisciano, E. R. Vyse and D. G. Cameron, *Wildlife Soc. Bull.*, 1991, **19**, 94.
10. K. S. Fairbrother, A. J. Hopwood, A. K. Lockley and R. G. Bardsley, *Meat Sci.*, 1998, **50**, 105.
11. S. E. Bartlett and W. S. Davidson, *Can J. Fish Aq. Sci.*, 1991, **38**, 309.
12. J. Quinteiro, C. G. Sotelo, H. Rehbein, S. E. Pryde, I. Medina, R. I. Peréz-Martín, M. Rey-Mendez and Mackie I.M., *J. Ag. Fd. Chem.*, 1998, **46**, 1662.
13. J. L. Ram, M. L. Ram and F. F. Baudoun, *J. Ag. Fd. Chem.*, 1996, **44**, 2460.
14. T. Matsunaga, K. Chikuni, R. Tanabe, S. Muroya and H. Nakai, *Proc. 43rd. Int. Conf. Meat Sci. and Technol.*, 1997, p319.

15. H. Rehbein, I. M. Mackie, S. E. Pryde, C. Gonzales-Sotelo, R. I. Peréz-Martín, J. Quinteiro and M. Rey-Mendez, *Inf. Fischwirtsh.*, 1995, **42**, 209.
16. I. M. Mackie, S. E. Pryde, C. Gonzales-Sotelo, I. Medina, R. I. Peréz-Martín, J. Quinteiro, M. Rey-Mendez and H. Rehbein, *Trends in Fd. Sci. and Technol.*, 1999, **10**, 9.
17. T. Matsunaga, K. Chikuni, R. Tanabe, S. Muroya, K. Shibata, J. Yamada and Y. Shinmura, *Meat Sci.*, 1999, **51**, 143.
18. K. S. Fairbrother, A. J. Hopwood, A. K. Lockley and R. G. Bardsley, *An. Biotech.*, 1998, **9**, 89.
19. A. J. Hopwood, K. S. Fairbrother, A. K. Lockley and R. G. Bardsley, *Meat Sci.*, 1999, **53**, 227.
20. C. R. Newton, A. Graham, L. E. Heptinsall, S. J. Powell, C. Summers, N. Kalsheker, J. C. Smith and A. F. Markham, *Nucl. Acids Res.*, 1989, **17**, 2503.
21. S. Rust, H. Funke and G. Assman, *Nucl. Acids Res.*, 1993, **21**, 3623.
22. A. K. Lockley, J. S. Bruce, S. J. Franklin and R. G. Bardsley, *Meat Sci.*, 1996, **43**, 93.
23. R. K. Saiki, P. S. Walsh, C. H. Levenson and H. A. Erlich, *Proc. Natl. Acad. Sci. USA*, 1989, **86**, 6230.
24. A. K. Lockley, C. G. Jones, J. S. Bruce, S. J. Franklin and R. G. Bardsley, *Nucl. Acids Res.*, 1997, **25**, 1313.
25. A. A. Oroskar, S-E. Rasmussen, H. N. Rasmussen, S. R. Rasmussen, B. M. Sullivan and A. Johansson, *Clin. Chem.* 1996, **42**, 1547.
26. S. P. A. Fodor, *Science*, 1997, **277**, 393.
27. D. P. Little, A. Braun, M. J. O'Donnell and H. Köster, *Nature Med.*, 1997, **3**, 1413.
28. S. Tyagai, D. Bratu and S. R. Kramer, *Nature Biotechnol.*, 1998, **16**, 49.
29. C. T. Wittwer, K. M. Ririe, R. V. Andrew, D. A. David and U. J. Balis, *Biotechniques*, 1997, **22**, 176.
30. I. A. Nazarenko, S. K. Bhatnagar and R. J. Hohman, *Nucl. Acids Res.*, 1997, **25**, 2516.
31. B. Kimura, S. Kawasaki, T. Fujii, J. Kusunoki, T. Hoh and S. J. A. Flood, *J. Food Prot.*, 1999, **62**, 329.

THE APPLICATION OF DNA BASED TECHNIQUES FOR THE DETERMINATION OF FOOD AUTHENTICITY

Jason Sawyer, Della Hunt, Neil Harris, Sally Gout, Clare Wood, Helen Gregory, David McDowell, Rita Barallon and Helen Parkes

Lifesciences Research Group,
LGC, Queens Road, Teddington, Middlesex ,TW11 0LY. U.K.

1 INTRODUCTION

Identification and detection of target DNA sequences which are specific for a species or sub-species can provide the basis for an unequivocal authenticity test. The detection of these DNA markers allows a variety of food authenticity issues to be addressed including areas as diverse as meat and fish species identification, the detection of genetically modified foods, wheat and rice varietal discrimination and beverage adulteration. In addition, there are other areas which have only recently begun to benefit from the use of DNA based techniques such as the specific identification of different animal tissues and sub-species and strain identification.

LGC's Lifesciences Research team has been awarded a number of major research projects over the last decade which have focused predominantly on the development and application of such techniques for the resolution of a number of food authenticity issues. Very significant progress has been made in demonstrating the usefulness of DNA in food authenticity. Our work is currently focused on several areas in which improvements of existing techniques would prove beneficial. These include:

- The analysis of processed and cooked foods where the extracted DNA can be highly degraded and contain PCR inhibitors
- Improvements in quantitative measurement
- Improvements in tests used to identify and distinguish samples likely to contain several components (such as species identification)

2 THE ANALYSIS OF PROCESSED AND COOKED FOODS WHERE THE EXTRACTED DNA CAN BE HIGHLY DEGRADED AND CONTAIN PCR INHIBITORS

The Polymerase Chain Reaction (PCR) is currently the preferred method for DNA detection for food analysis. It is the most appropriate technique because it enables the sensitive detection of small amounts of target DNA. This allows the determination of the relative amounts of food components even when they are present at very low levels in a finished product. However, we regularly find DNA recovered from food samples to be degraded by the actions of heat and the chemical / mechanical processes occurring during food processing resulting in the recovery of fragments that can range from a high molecular weight to that of only a few hundred base pairs or less. In addition, degradation reduces the number of intact DNA targets present in the final preparation (even though the total amount of DNA recovered may be the same). This may affect the accuracy of quantitation and emphasises the requirement for the use of appropriate controls and interpretation.

To minimise the effects of DNA degradation and ensure that as many types of product as possible can be tested using a particular primer pair we aim to be able to detect increasingly smaller sections of DNA. By designing primer sequences to amplify a target of less than 200 bp and preferably 80-120 bp, it has been possible to carry out amplification of DNA from a range of highly processed sources. Such examples are the detection of GM ingredients where we have identified target material in a range of sources from flour (Figure 1) to processed ready meals and confectionery items (Figure 2) [1] and the determination of meat speciation in tinned and pre-cooked products.

Individual food constituents (such as carbohydrates, tannins, fats) will also act as PCR inhibitors if they are not completely removed by the DNA extraction method employed or enhancers are not added to the amplification reaction. [2,7] Several techniques are available ranging from crude DNA preparations to complex multistep protocols. Several approaches have been evaluated at LGC during the course of our research. The most suitable method, applicable to a wide of starting materials, has been found to be homogenisation followed by proteinase K digestion, of a large sample of the product (to obtain a representative sample) then purification of a sub-sample with a commercial purification column (Wizard DNA Clean-up system, Promega).

3 IMPROVEMENTS IN QUANTITATIVE MEASUREMENT

One of the most challenging areas for food authenticity techniques is the generation of accurate quantitative measurements. It is essential for enforcement of legislation that quantitative measurement is possible, given that legislation often sets out the levels of materials in products that constitute adulteration, as opposed to adventitious contamination and permitted levels of ingredients. Quantitative measurement with PCR is technically more difficult and challenging than using more traditional hybridisation techniques and our research has focused on overcoming these problems.

Previous work in this area at LGC involved amplification of an internal mimic alongside the target DNA from the unknown sample. This has focused on developing model systems to evaluate the critical parameters involved in quantitation using this technique[3]. We have found that amplicon length and GC content should be matched as near as possible to avoid differential amplification. In addition, the effects of island motifs, which can serve as termination sites leading to differential amplification, can be minimised by the use of co-solvents such as betaine or proof reading enzymes such as Pfu polymerase. The mimic contains the same PCR priming sites as the target DNA in the unknown sample and acts as a competitor in the amplification process. This system can be set up so that the relative ratio of PCR products derived from the mimic and the target DNA are indicative of the original concentration of target DNA in an unknown sample. Because the mimic is included as an internal control in the same tube as the unknown sample DNA a greater degree of accuracy is obtained than that of limits of detection analysis.

Recent technical developments, enabling measurement of the PCR process in real time, have provided an alternative system for obtaining quantitative measurements. Current real time amplification/detection systems such as TaqMan (Perkin Elmer) and Lightcycler (Roche) use direct measurement of an increase in emitted fluorescence as PCR products are synthesised. At LGC, we have focused on using the latter system to carry out quantitative measurements and potentially replace gel based analysis. The lack of gel analysis, coupled with the high rate of amplification of these systems, greatly reduces the time needed for analysis. Real time monitoring is extremely sensitive and allows the identification of the stage in the PCR cycle at which products are first amplified to a detectable level. The higher the amount of target sequence the sooner this will happen in a typical PCR reaction. Detection at such an early stage in the amplification process ensures that reactions are monitored prior to any potential inhibitory factors having an effect upon amplification efficiency, increasing the inherent accuracy of the data obtained. By comparing the first appearance of amplified product from an unknown sample to that of a range of known standards we are able to calculate the amount of target sequence in the sample.

The use of real time PCR also provides a potential solution to the inaccuracies caused by differences in the levels of DNA degradation between an unknown sample and known standards. A control PCR primer set (amplifying a common DNA target) can be used to measure the reduction in signal intensity that results from degradation of the sample and this is taken into account when measuring the signal obtained from the adulterant specific signal. The ratio of the signal obtained for the adulterant specific and control reactions should be constant irrespective of the degradation state of the samples. One area where this has been employed is in the detection of GM soya where the total soya content (non-GM plus GM) is measured by the detection of a soya lectin gene and this is compared to the signal generated from GM Soya specific amplification for a particular sample (Figure 3).

4 IMPROVEMENTS IN TESTS USED TO IDENTIFY AND DISTINGUISH SAMPLES LIKELY TO CONTAIN SEVERAL COMPONENTS (SUCH AS SPECIES IDENTIFICATION)

Because of the large number of species potentially encountered in species testing a 'universal' approach such as CAPS (Figure 4) or SSCP analysis (where all species can be PCR amplified and further analysed to reveal species specific profiles) is highly attractive and these technique have, to date, been the method of choice for this type of analysis. [4,5]

We have found that CAPS analysis, however, can be limited by the potential differential amplification of species in mixed samples (which can result in false under or over representation of species in the results), the increasing number of enzymes required to distinguish large numbers of species (which results in expensive and cumbersome experimental analysis) and the complex, difficult to interpret banding patterns which result from mixed samples. For SSCP analysis, the technique is time consuming and substantial experience is needed to achieve reproducible results. In both cases, because of the non-specific initial PCR step involved there is the potential for differential amplification of species in mixed samples.

To circumvent these problems our research on meat species identification has focused on the development of a one species:one test approach. Such tests were originally based upon the use of Southern hybridisation probes but we have now developed a set of species specific PCR primers for use on a wide range of meat products. These are being used in conjunction with real time PCR detection in the LightCycler. This will allow rapid accurate quantitation of the relative levels of each constituent species in a mixed sample and bypass the need for gel based visualisation of results that CAPS and SSCP analysis rely on.

5 NOVEL APPLICATIONS OF DNA TECHNIQUES FOR FOOD AUTHENTICITY TESTING (How about using them as toilet paper?)

5.1 Tissue specific identity in meat products - bone marrow (MRM) in chicken and neuronal tissue in bovine, sheep and goat products

Research at LGC over the last two years has resulted in the development of an innovative PCR based assay for distinguishing different tissues from the same species. This is being exploited to identify bone marrow, which acts as an indicator for the presence of mechanically recovered meat (MRM) in chicken and neuronal tissue in bovine, sheep and goat products.

The assay is based on the identification of the control elements of genes that are expressed in bone marrow or neuronal tissue but not in muscle. Primers are then designed to detect these elements by PCR amplification. The technique developed will only amplify the target genes from DNA derived from neuronal or bone marrow tissue and not from muscle tissue, indicating the presence of these tissues in a given sample. We are currently developing this technique into a robust assay system and have demonstrated that it is applicable to both raw and cooked samples.

5.2 Variety identification by genetic profiling - AFLP analysis [6]

AFLP (Amplified Fragment Length Polymorphism) profiling is a powerful technique that generates a highly discriminatory genetic profile allowing differentiation between closely related organisms at the sub-species and varietal level. LGC is currently using this technique to rapidly identify strains of *E. coli* 0157 and *Salmonella*, something that is difficult with traditional biochemical and microbiological techniques. This information is being used to set up a database to allow the rapid characterisation of samples collected from suspected outbreaks of these pathogens.

6 FUTURE DEVELOPMENTS

Existing techniques still rely on a PCR amplification step but additional technologies are emerging that have the potential for direct detection. DNA based food analysis is likely to benefit from the advances in detection technology developed by the medical diagnostics field, an area of current research at LGC. Such techniques include the use of fluorescent probes, DNA arrays and microchips. Such methods offer the possibility of being able to carry out many measurements simultaneously on a small affordable scale.

7 CONCLUSIONS

At LGC we have made very significant progress in demonstrating the feasibility of DNA approaches to solve a variety of challenging food authenticity problems. Such DNA based methods offer significant advantages over existing technologies, particularly when applied to cooked and processed foodstuffs. However, there remain validation issues to be resolved as the new tests, particularly quantitative assays, move from the research laboratory to routine application [7]. In addition, a range of new analytical techniques, many of which no longer rely on a PCR amplification step, need to be investigated for their applicability to future food authenticity analysis and these will be the subject of continued research at LGC.

1. H.C. Parkes, *Chemistry in Britain*, 1999, **35**, 32.
2. J. Bickley, J.K. Short, D.G. McDowell and H.C Parkes, 1995, *Lett. Appl. Microbiol.*, **22**, 153.
3. D.G. McDowell, N.A Burns and H.C Parkes, 1998, *Nuc. Acids Res.*, **26**, 3340.
4. R. Meyer, C. Höfelein, J. Lüthy and U. Candrian, *J. AOAC Int.*, 1995, **78**, 1542.
5. R. Barallon, 'Forensic DNA protocols' (eds. P.J Lincoln and J. Thompson) Humana Press, New Jersey, US, 1998, p251.
6. P. Vos, R. Rogers, M. Bleeker, M. Reijans, T. van der Lee, M. Hornes, A. Frijters, J. Pot, J. Peleman, M. Kuiper and M. Zabeau, *Nuc. Acids. Res.*, **23**, 4407.
7. G.C Saunders and H.C. Parkes (eds), 'Analytical Molecular Biology: Quality and Validation' Royal Society of Chemistry, Cambridge UK, 1999.

TESTING FOR GENETICALLY MODIFIED COMPONENTS IN FOODS - PRESENT AND FUTURE CHALLENGES

A. B. Hanley, H. Hird and M. L. Johnson

Central Science Laboratory, Sand Hutton, York YO41 1LZ, UK

1   INTRODUCTION

The requirement for testing for genetically modified (GM) components in foods is both legislatively and consumer driven. In September 1998, new EU legislation (Novel Food Regulation 258/7 and Council Regulation 1139/98) came into force which stated that novel foods and food ingredients which contain recombinant DNA or modified protein classifies them into the 'no longer equivalent' category, and hence, will have to be labelled, although the threshold at which labelling must occur is yet to be agreed. Consumer concerns have been increasing over recent months and there is considerable interest in labelling for consumers' benefit.   These have provided considerable challenges for current methods while new advances will provide future problems.

2   TESTING FOR GENETICALLY MODIFIED COMPONENTS IN FOODS - PRESENT CHALLENGES

2.1   Background

The use of GM components in foods has a lengthy history but it is only in the relatively recent past that GM plants have been developed for human food uses, those currently licensed for sale around the world include tomato, potato, cotton, sugar beet, tobacco and rape seed[1]. By far the largest source of genetically modified foods are those containing soya or maize. The increase in size of the GM crop has been substantial with, for example, an anticipated 50% of the US soya harvest expected to be GM in 1999.

The principal modification for soya and maize are respectively the insertion of the petunia gene, EPSPS, which confers resistance to the Monsanto herbicide, Roundup Ready and the insertion of a bacterial gene, Bt, which confers resistance to the corn borer. The detection of GM maize and soya, particularly in convenience and processed foods, provides the major current challenge for GM testing.

There are two ways in which genetic modifications in foods can be detected: by protein based methods or DNA based methods, each has limitations but are applicable under different circumstances.

2.2   Methodology

Protein-based methods of testing for GM foods rely on the specificity and selectivity of antibodies. These methods have significant limitations. Raising antibodies to the proteins resulting from genetic modification is difficult and can be laborious. More critically, proteins degrade on processing, changing the regions to which antibodies bind, causing a loss of antibody reactivity. Therefore, protein based methods are generally only appropriate to unprocessed food or starting materials.

DNA based methods are currently used by most commercial laboratories for the detection of GM soya and maize in processed foods. The "foreign" DNA introduced during genetic engineering can act as markers for GM plants. Markers include:

- The new gene introduced to elicit the required characteristic
- The promoter and terminator sequences which flank the introduced gene and act as molecular switches to ensure that the introduced gene functions properly
- Antibiotic resistance genes introduced to aid selection and development of the GM plant

The first step in the detection of GM food is extraction of DNA from the food sample. A fragment of the DNA which comprises some of the inserted gene is then copied many millions of times, using the polymerase chain reaction (PCR). The DNA amplicon is analysed by agarose gel electrophoresis, where the presence of amplicons of the correct size indicate genetic modification. Detection limits are in the range of 20pg to 10ng of target DNA or 0.0001% to 1% GM in non-GM material. Currently the majority of commercially released GM crops can be identified using a limited number of target fragments although more specific assays, in which the introduced gene is targeted can also be used.

2.3   Current drawbacks

Currently GM components can only be detected in processed foods using DNA based methods but the extent of food processing may give rise to variability in GM detection. For example, it is possible to extract DNA from raw pressed oils, but not from highly refined oils, and GM material can be detected in some maize starch samples but not in others, possibly due to interbatch variability in processing. The main reasons for this variability is that DNA breaks up under heat and pressure treatment [2]. PCR depends on a fragment of DNA being left intact after processing, however various chemical, physical and enzymatic factors contribute to DNA degradation in processed foods:

- Prolonged heat treatments may result in DNA hydrolysis, or may modify the structure of the DNA in such a way that the PCR process may not work

- Increased chemical modification and hydrolysis of DNA at low pH (eg vinegar)
- Enzymatic degradation of DNA by endogenous nucleases may occur on prolonged storage of fresh foods

In addition the presence of PCR inhibitors in foodstuffs including cations, eg. Fe^{3+}, Ca^{2+}, trace heavy metals, carbohydrates, tannins, phenolics and salts, eg. sodium chloride and nitrites can reduce the efficiency of PCR.

Both DNA degradation and the presence of PCR inhibitors can result in false negative results when testing for the presence of GM components in foods. These problems can, however, be minimised by careful primer design and stringent DNA isolation conditions. The average length of fragmented DNA for PCR must be at least the length of the amplicon for PCR to be efficient. Primers are therefore designed to amplify small fragments of DNA. Many of the commercially available primers amplify fragments around 100 base pairs, which is usually of a sufficiently small size for the target to still be present in processed foods. The amplification problems accompanying PCR inhibitors can be minimised with the use of state of the art DNA isolation technology, for example DNA binding resin allows recovery of very pure DNA. Testing for GM components in foods, involves not only PCR for the foreign DNA, but must also include positive control reactions with primers designed to react to any plant material, providing information on template quality and the presence of inhibitors, and negative control reactions, providing information on the quality control of the testing laboratory.

3 FUTURE CHALLENGES TO TESTING FOR GMO

3.1 Background

New legislation will shortly agree a threshold level for GM food in food products, above which there will be a requirement for declaration. In response, greater efforts will be needed to be made to develop quantitative GMO testing.

3.2 Quantitation

Current analytical methods for detecting GM food are essentially semi-quantitative, and largely based on limits of detection compared with reference standards. More accurate methods for quantifying GM materials in foods are "competitive" PCR, in which the amplification of the GM target fragment is monitored relative to an internal reference target and true quantitative PCR with Taqman™ technology[3].

Taqman technology relies on the release of a fluorescent reporter molecule after each PCR cycle. The more DNA template present, the sooner fluorescence is detected. Taqman technology is particularly efficient at detecting DNA fragments of between 60 and 80 base pairs and is therefore uniquely placed to detect GMO in processed food products where DNA may be highly degraded.

Genetically modified plants are now being developed which have more than one genetic modification. Current PCR technology is unable to distinguish between these doubly transformed plants and mixtures of plant lines containing the transformations singly, for example, maize which has been transformed with both the Bt and EPSPS genes will be indistinguishable compared to a mixture of two maize plants transformed with Bt or

EPSPS. Using Taqman technology the exact quantities of each gene could be calculated and the presence of either the doubly transformed lines or the two singly transformed lines determined. This is an important point for labelling regulations when threshold values have been agreed. If a food product contains one GM plant with two genes, it may fall below the threshold value, however if the same food product contains two plant lines genetically modified with either, but not both genes then these may have to be additively declared under the labelling regulations, even though the quantity of 'foreign' DNA would be the same.

Taqman has moved PCR technology from qualitative and semi-quantitative to fully quantitative and provides any testing service with an excellent tool for detecting GM components in food products.

3.3   Protein based methods

DNA based technology is ideal in certain situations however, protein based approaches are often easier to implement and indeed, may prove to be crucial in the traceability of GM and non-GM crops. Products which contain neither DNA nor protein are outwith the realms of the current legislation. These include sugars, hydrolysed starch products and refined oils. The validation of these crops, from the seed before it is planted, to transport and subsequent processing is the only way in which checks on the GM status of the crops can be monitored.

The most promising protein based assays are currently based on enzyme immunoassay (EIA) technology which has been shown to be an accurate, sensitive and simple method to detect specific molecules in a huge range of materials, including soil, blood, food products and in both plant and animal tissues.

For some years now EIA technology has been used to develop dipstick assays which are beginning to be available for GM testing. Using dipstick assays the fate of GM and nonGM crops can be monitored by the farmer, the wholesaler and the processor, ensuring traceability and validation of the crops. Dipstick assays are typically sensitive (0.1% threshold), quick, (3-5 mins) and provide a definitive answer with little or no interpretation. The challenge however will be to develop these dipsticks at a pace in line with the GM food development, and since immunoassay production is a long process, it may significantly lag behind the increasing array of crop plants undergoing genetic modification.

3.4   Recent innovations

One of the major problems with GMO foods is the public perception of 'foreign' DNA introduced into the genome of the host species. 'Foreign' DNA usually includes the target gene and a complement of other sequences, ensuring the correct function of the gene in its new setting. In an effort to allay public fears, efforts are being made to subtly alter the sequence of endogenous genes to create proteins with new functions. This technique, referred to as chimeraplasty, has been successfully used on tobacco plants to confer resistance to sulphonylurea weedkillers [4]. The DNA of the host plant was altered so that the product of the altered gene had only one amino acid substitution. The remainder of the host DNA was unaffected and was therefore indistinguishable from non-GM plant DNA. The present practice of inserting a relatively large 'foreign' gene allows great scope in the design of PCR primers and a completely new protein to act as antigen for antibody production, however chimeraplasty, where only 3 bases or one amino acid may be altered, severely limits primer design and calls for exceptionally well optimised PCR conditions or antibody selection, and may well lead to false negative results.

4   CONCLUSION

The importance of validated, sensitive, quantitative and specific methods for testing GM foods is clear to consumers, legislators and the scientific community. The advances made in GM technology for crop plants will continue to act as a spur for the development of newer and more sensitive methodologies which will contribute to the growing array of techniques for GM detection.

1. E. Gachet, G.G. Martin, F. Vigneau and G. Meyer, *Trends in Fd. Sci. and Technol.,* 1999, **9**, 380.
2. R. Meyer, U. Candrian and J. Luthy, *J. AOAC Int.,* 1994, **77**,617.
3. Y. S. Lie and C.J. Petropoulos *C. Op. Biotech.,* 1998, **9**, 43.
4. P. R. Beetham, P. B. Kipp, X.L. Sawycky, C. Arntzen and G.D. May, *Proc. Natl. Acad. Sci.,* 1999, **96**, 8774.

TOXICOLOGICAL END-POINTS - THE ZERO RABBIT OPTION

E. J. Hughson, S. L. Oehlschlager, and A. B. Hanley

Food Safety and Quality Group
Central Science Laboratory
Norwich NR4 7UQ.

1 INTRODUCTION

Food constituents which are ingested in the diet not only have nutritional value but may also have beneficial or adverse effects on human health. For example, it has been estimated that 20-40% of cancers are related to dietary exposure.[1] There are stringent requirements for testing the potential toxicity of dietary constituents and these assays are usually *in vivo* in nature. The aims of this paper are to discuss why the use of in vitro systems should be considered and to look briefly at some examples which have been developed by the authors .

1.1 The Advantages of *in vitro* Systems

1.1.1 Unsuitability of animals as model systems. In many instances, there are important interspecies differences in biochemical, physiological and metabolic processes which could invalidate results from tests conducted in animals. A current area of investigation is the cytochrome P450 isotypes which appear to show some interspecies differences in specificity for their substrates (for example, see reference 2). This could have important implications for studies on the metabolism of possible toxic compounds conducted *in vivo.*

1.1.2 Speed, capacity and sensitivity of in vitro systems. When food compounds are being developed, *in vitro* assays offer the potential for multiple screening. Such systems may be suitable for automation and in many cases will be rapid in comparison to *in vivo* studies. When only limited quantities of test compounds are available, smaller amounts should be required given the innate sensitivity of *in vitro* systems. For example, *in vivo* studies for testing the allergenicity of proteins require milligram quantities of material,[3] whereas an *in vitro* system which is in development (see below and reference 4) uses microgram quantities.

1.1.3. Ethical considerations. In an ideal world, testing of both dietary and pharmaceutical compounds on animals would be minimal or unnecessary. Public opinion has given momentum to the withdrawal of cosmetics testing in animals in the U.K. Newer technology is developing which could maintain this momentum into other areas; an example is "Apigraft", the artificial "skin" currently used for wound healing being marketed by Organogensis Inc. [5]

1.1.4. Understanding toxicological mechanisms. When dietary constituents are tested in whole animals, it can be difficult to elucidate the mechanisms mediating toxic effects and beneficial effects could be missed. In this case *in vitro* systems are essential for further investigation, particularly in research and development projects. Cellular based systems are both easier to standardise and to manipulate experimentally.

However, there are as yet no alternatives to *in vivo* testing when it is necessary to examine systemic effects such as neurotoxicity.

2 EXAMPLES OF *IN VITRO* ASSAY SYSTEMS

2.1 The choice of experimental material

The types of material needed for *in vitro* assays can be broadly divided into two types, primary tissue and secondary cell lines (cell-free systems will not be considered here). The latter have the advantages of being readily available in large quantities, last a relatively long time and are usually homogeneous. However, such cells will have lost many of the differentiated characteristics of the parent tissue and may show some features typical of transformation.

Two *in vitro* systems will be considered; the first, an intestinal cell line which has been used for a number of years and the second, an *in vitro* system to test for possible allergens using primary tissue which is being developed in the authors' laboratory.

2.1 Cell lines

A number of human intestinal cell lines have been developed, including the Caco-2 cell line which was derived from a human colon carcinoma.[6] When grown on a permeable filter support, these cells can become highly polarised and show many characteristics of small intestinal enterocytes, including a well-developed brush

Table 1 *Diet-related applications of Caco-2 cells*

Application	Reference
Peptide transporter studies	7
Sugar transporter studies	8,9
Assessment of bacteria binding	10
Heavy metal toxicity	11
Immunological function of the gastrointestinal epithelium	12
DNA damage caused by genotoxins	13

border, apical membrane glycolases and peptidases, and secretion of typical proteins. This cell line has been used for a multitude of purposes (Table 1), illustrating the versatility of such systems.

2.2 Primary tissue - an *in vitro* system for the assay of potential allergens

2.2.1. The allergic response. Food allergy, as distinct from food intolerance and food avoidance, has an immunological basis. In the following sections, IgE-mediated responses only will be considered (for a detailed explanation, see reference 14). A sub-population of the antibody synthesising cells or B lymphocytes, for reasons which are not yet clearly understood, are directed to produce antigen-specific IgE, the antibody isotype which mediates allergic reactions. These antibodies circulate in the blood and bind to specific receptors on the surface of mast cells at mucosal surfaces such as the gastrointestinal tract. In atopic (ie allergy-prone) individuals, when an appropriate antigen with multiple antibody binding sites or epitopes is encountered, the receptor-bound IgE antibodies bind to the epitopes, causing the receptors to cluster and cross-link. This signals to the mast cell to degranulate and release molecules such as tumour necrosis factor alpha (TNF_α) which mediate the allergic response.

2.2.2. The need for assays to detect allergens. In recent years, there has been an increasing awareness of allergy to naturally occurring food components. In addition, developments in food processing such as microparticulation, the emergence of genetically modified food technology and the introduction of new foods into the diet could lead to the exposure of hidden allergenic epitopes or the creation of novel epitopes. Thus there is a need for assays to screen food components and food complexes for possible allergenicity. Animal models have been developed, but have the disadvantage that conditions can be manipulated so that an IgE response can be produced against almost any protein. Existing human tests in the human such as the RAST (radioallergosorbent test) assay, although detecting the presence of specific IgE, do not show whether an allergic reaction will take place or not.

2.2.3 Development of an in vitro system for the assay of potential allergens. An experimental assay system to detect human mast cell responses to allergen is being developed in our laboratory.[4] The study was designed to detect known existing allergens and molecules containing cross-reactive epitopes. Due to the source of the IgE (sera from allergic individuals, kindly provided by the Norfolk and Norwich Allergy testing service), it is unlikely that novel epitopes will be recognised. Fresh human lung tissue containing mast cells (Norfolk and Norwich General Hospital) is minced, aliquoted and incubated overnight with allergenic sera to allow IgE to bind to the mast cell receptors. The tissue is incubated the following day for 4 hours with allergen, and the medium is analysed by ELISA to monitor the release of TNF_α. Characterisation of the system has shown the following:

a) in the presence of allergen, an IgE-dependent release of TNF_α is obtained,

b) the effect can be mimicked by replacing the allergen with an antibody against IgE, which because of its bivalent nature cross-links the IgE antibodies (and hence mast cell receptors)

c) in the presence of a constant amount of sera, there is a greater release of TNF_α with smaller tissue aliquots (hence less mast cells per aliquot). This is with consistent with the concept that when more IgE receptors per cell occupied, the greater the potential for clustering.

d) when TNF_α is titrated against concentration of allergen, at least for some allergens such as Timothy grass allergen,[4] a peak of response is obtained and as the concentration increases further, the response declines. This suggests the allergen is now in excess, resulting in a decrease in cross-linking.

This assay thus has the potential to test for the presence of known and cross-reacting allergenic epitopes in food components.

3 CONCLUSION

It will be seen from above that *in vitro* systems have certain advantages over animal model systems, but for the foreseeable future, *in vitro* systems will provide complimentary information to animal models, rather than offer a complete alternative. For an in depth dicussion on alternatives to tests on food chemicals in animals, the reader is referred to reference 15.

Acknowlededegements
We would like to thank the Ministry of Agriculture, Fisheries and Food for their support.

References
1. B.N Ames., L.S. Gold and W.C. Willett,. *P.N.A.S.*, 1995, **92**, 5258.
2. B.J. Brock and M.R. Waterman, *Biochemistry,* **38**,1598.
3. L.M. Knippels, A.H. Penninks, S. Spanhaak and G.F.Houben, *Clin. Exp. Allergy*, 1998, **28,** 368.
4. E.I. Opara, S.L. Oehlschlager, L. Day, S.A. Ridley, R.Vaughan, and A.B Hanley,., *Toxicology in Vitro,* in press.
5. W.M. Saltzman, Nature Medicine, 1998, **4**, 272.
6. Pinto, S. Robine-Leon, M. Appay, M. Kedinger, N. Triadou, E. Dussaulx, B. 7. Lacroix, P. Simon-Assmann, K. Haffen, J. Fogh and A. Zweibaum, *Biol. Cell*, 1983, **47,** 323.
7. K. Tamura, K.A. Agrios, D.V. Velde, J. Aube and R.T. Borchardt, *Bio. Med, Chem.*, 1997, **5**, 1859.
8. A. Blais, P. Bissonnette and A. Berteloot, *J. Membrane Biol.*, 1987, **99,** 113.
9. D.S. Harris, J.W. Slot, H.J. Geuze and D.J. James, P.N.A.S., 1992, **89**, 7556.
10. M. Bernet, D. Brassart, J. Neeser, A.L. Servin, *Appl. Environ. Microbiol.,* 1993, **59**, 4121.
11. A. Rossi, R. Poverino, G. Di Lullo, A. Modesti, A Modica and M.L. Scarino, *Toxicology in Vitro,* 1996, **10**, 27.
12. L. Lopes, E.Hughson, Q. Anstee, D. O'Neil, D. Katz and B. Chain, *Immunol.* (in press).
13. M. Venturi, R.J. Hambley, B. Glinghammer, J.J. Rafter, and I.R. Rowland, *Carcinogenesis*, 1997, **18,** 2353.
14. J.Kuby, 'Immunology', W.H. Freeman and Co., New York, 1997.
15. K. Botrill, *ATLA*, 1998, **26**, 421.

FAPAS®: AN INDEPENDENT ASSESSMENT OF LABORATORY PROFICIENCY

A. L. Patey

FAPAS® Secretariat
CSL Food Science Laboratory
Colney Lane
Norwich NR4 7UQ
United Kingdom

1 INTRODUCTION

The Food Analysis Performance Assessment Scheme (FAPAS®) received Ministerial authority to start operation in 1990. Previous to this the UK Ministry of Agriculture, Fisheries and Food (MAFF) had some evidence that surveillance data on contamination of foodstuffs was not adequate for its purposes and thus, MAFF required a Scheme to assess the competence of laboratories. It was expected that about 50 UK laboratories would be taking part in FAPAS® which would run eight series of different test materials. It was predicted two staff would be required to run the Scheme.

However, once FAPAS® started it soon became a very popular Scheme and its reputation for speedy and fair assessment of analytical performance spread widely. Laboratories in many Countries that did not have a scheme of their own requested participation and there were many requests to increase the breadth of testing, for example into areas such as migration testing and feedingstuffs. Thus, the Scheme was expanded. Today, 200 laboratories situated in the UK take part together with 400 laboratories in 56 other Countries. Twenty Series of test materials are issued on a regular basis and twelve staff are required to run the Scheme. In the last 12 months over 5000 test materials have been distributed and since 1990, over 70,000 proficiency assessments made, with 81% satisfactory.

2 SCHEME PROTOCOL

Before FAPAS® could start its job of assessment a protocol had to be devised on how the Scheme was to operate. In 1989 an Advisory Committee was formed by MAFF to do this. The Committee was formed from senior figures in the United Kingdom food analysis community. (It was subsequently enlarged to include EU and feed analysis representatives). The Committee decided that performance would be assessed by a z-Score procedure. The z-score compares the assigned analyte value ($\hat{X}$) with the participants answer (x) and divides by an externally-derived standard deviation term (σ) to give the equation:

$$z = \left(\hat{X} - x\right)/\sigma$$

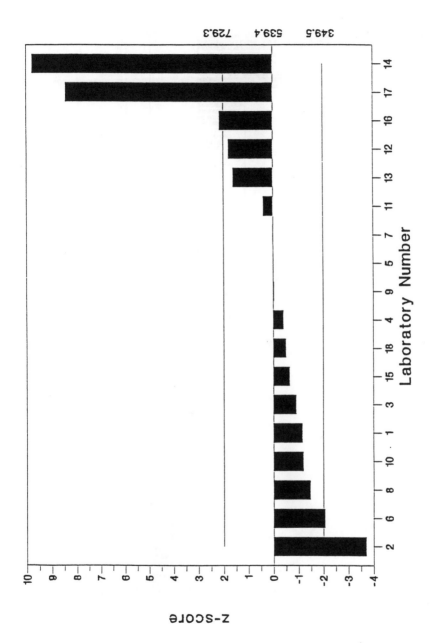

Figure 1 *z-Score for sulfamidine (539.4 µg/kg)*

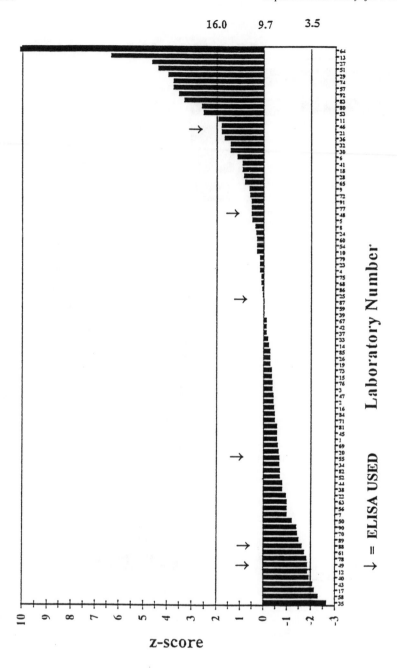

Figure 2 *z-Scores for total aflatoxins (9.7 µg/kg) in paprika test material*

z-Scores between +2 and -2 are deemed satisfactory. Those over +3 or below -3 are unsatisfactory. The z-score procedure is easy to understand and apply and readily allows analysts, managers and assessors a clear picture of the performance of a laboratory.

The Committee also decided that analytical methods used by laboratories would not be specified and laboratories could use the method of analysis of their choice. Assigned values would be the robust mean of all data supplied, but not considering "impossible" results (eg 1 kg/100g). Thus participants could use classical and well-established methods of analysis or choose a rapid detection assay, perhaps one that had just been designed. Although FAPAS® was not instigated to compare methods of analysis it could be used for this purpose. The majority of rapid methods used have been in the analysis of veterinary drug residues and aflatoxin analysis, although it must be said that up to the present these have been in a minority.

3 ASSESSMENT OF RAPID MATTERS

In a Round of veterinary drug residue testing, the test material, pig muscle, contained sulfadimidine. 18 Laboratories took part and each gave a brief description of the method used along with their result for the amount of drug present. 16 Participants used a range of HPLC or TLC methods with a diverse range of extractions, but two laboratories (Numbers 17 and 14, See Figure 1) stated they used ELISA techniques. These two laboratories grossly over-estimated the amount of drug present. This was a disappointing result for the use of new rapid methods.

However my second example paints a different picture. The test material was paprika and 92 laboratories took part. 86 Participants used a range of HPLC or TLC methods and 6 laboratories used ELISA methods. All 6 of these participants produced a satisfactory result for the amount of aflatoxin present, whilst 15 of those that used HPLC/TLC methods were not in the satisfactory z-score range (see Figure 2). This was just the opposite of the first example.

4 CONCLUSION

Because participants in FAPAS® can use the method of analysis of their choice there is an opportunity to be able to compare the performance of newer rapid methods of analysis with more traditional (slower) methods. Reliability and accuracy of methods can be assessed. Although not the prime purpose of FAPAS®, Scheme data can be used for this purpose for independent assessment of laboratory and method proficiency.

Subject Index